NOAH K.

GOD'S DESIGN® FOR HEAVEN & EARTH

OUR UNIVERSE

Noah J. K.
Awesomeness

project due 15th

1:1
answersingenesis
Petersburg, Kentucky, USA

3RD EDITION | UPDATED, EXPANDED & FULL COLOR

ANSWERS IN GENESIS SCIENCE BY DEBBIE & RICHARD LAWRENCE

God's Design® for Heaven & Earth is a complete earth science curriculum for grades 1–8. The books in this series are designed for use in the Christian homeschool and Christian school, and provide easy-to-use lessons that will encourage children to see God's hand in everything around them.

Third edition
Fourth printing February 2012

Copyright © 2008 by Debbie and Richard Lawrence

ISBN: 1-60092-153-1

Cover design: Brandie Lucas & Diane King
Interior layout: Diane King
Editors: Lori Jaworski, Gary Vaterlaus

The publisher and authors have made every reasonable effort to ensure that the activities recommended in this book are safe when performed as instructed but assume no responsibility for any damage caused or sustained while conducting the experiments and activities. It is the parents', guardians', and/or teachers' responsibility to supervise all recommended activities.

Published by Answers in Genesis, 2800 Bullittsburg Church Rd., Petersburg KY 41080

Printed in China

PHOTO CREDITS

TABLE OF CONTENTS

WELCOME TO
GOD'S DESIGN®
FOR HEAVEN & EARTH

You are about to start an exciting series of lessons on earth science. *God's Design® for Heaven and Earth* consists of three books: *Our Universe*, *Our Planet Earth*, and *Our Weather and Water*. Each of these books will give you insight into how God designed and created our world and the universe in which we live.

No matter what grade you are in, first through eighth grade, you can use this book.

1st–2nd grade

Read only the "Beginner" section of each lesson, answer the questions at the end of that section, and then do the activity in the ████ box (the worksheets will be provided by your teacher).

3rd–5th grade

Skip the "Beginner" section and read the regular part of the lesson. After you read the lesson, do the activity in the ████ box and test your understanding by answering the questions in the ████ box.

6th–8th grade

Skip the "Beginner" section and read the regular part of the lesson. After you read the lesson, do the activity in the ████ box and test your understanding by answering the questions in the ████ box. Also do the "Challenge" section in the ████ box. This part of the lesson will challenge you to go beyond just elementary knowledge and do more advanced activities and learn additional interesting information.

Everyone should read the Special Features and do the final project. There are also unit quizzes and a final test to take.

Throughout this book you will see special icons like the one to the right. These icons tell you how the information in the lessons fit into the Seven C's of History: Creation, Corruption, Catastrophe, Confusion, Christ, Cross, Consummation. Your teacher will explain these to you.

Let's get started learning about God's design of our amazing universe!

UNIT 1

SPACE MODELS & TOOLS

INTRODUCTION TO ASTRONOMY

Study of space

LESSON
1

What is astronomy and why should we study it?

Words to know:

astronomy

big bang theory

BEGINNERS

When you look to the sky during the day, what is the brightest thing you see? The sun, of course. But at night, you do not see the sun. Instead, you see the moon and the stars. The planet we live on, earth, is part of something bigger than just this planet. The earth is just one planet in our solar system, and the sun is just one star in the universe. The Bible tells us that the sun, moon, and stars were created by God on the fourth day of creation.

We are getting ready to learn all about the universe and the things that God has put there. This study is called astronomy, and you are going to learn all about the planets, the sun, the moon, the stars, and much more. So get ready to learn about our universe.

- **What is astronomy?**

Psalm 19:1 says, "The heavens declare the glory of God." Since the Bible is always true, we should want to study and understand the heavens so that we can better understand God's glory. The study of the heavens is called astronomy. **Astronomy** is the study of the planets, moons, stars, and other things found outside of the earth. In this book you will learn about these things and many other things in the universe as well.

Have you ever looked at the stars and wondered what they were or how they got where they are? Have you ever observed the movement of the sun through the sky and wondered how it moves like it does? Then you are asking some of the same questions that astronomers have asked for hundreds of years. Scientists cannot prove where the universe came from. A popular theory among some scientists is the **big bang theory**—that all that exists in our universe came from a cosmic explosion about 14 billion years ago. However, in Genesis 1:14–19, the Bible says that God created the sun, moon, and stars on the fourth day of creation, so we know how the sun, moon, and stars got where they are—God created them. Many of the other questions have been answered by scientists as they have observed the universe and studied how things move and work together. In this book, you will learn many of the things that astronomers and other scientists have discovered as well as many things that the Bible has to say about the universe that we live in. ■

GOD'S PURPOSE

Complete the "God's Purpose for the Universe" worksheet.

WHAT DID WE LEARN?

- What is astronomy?
- Why should we want to study astronomy?

TAKING IT FURTHER

- What is one thing you really want to learn during this study?
- Write your question or questions on a piece of paper and save it to make sure you find the answers by the end of the book.

KNOWLEDGE OF THE STARS

Astronomy comes from two Greek words, which mean "knowledge of the stars." What knowledge do you have of the stars? Test your knowledge of the stars by completing the "Knowledge of the Stars" worksheet. Try to find the answers to the questions you are not sure of in any books you may have on astronomy or on the internet.

SPACE MODELS

What's really out there?

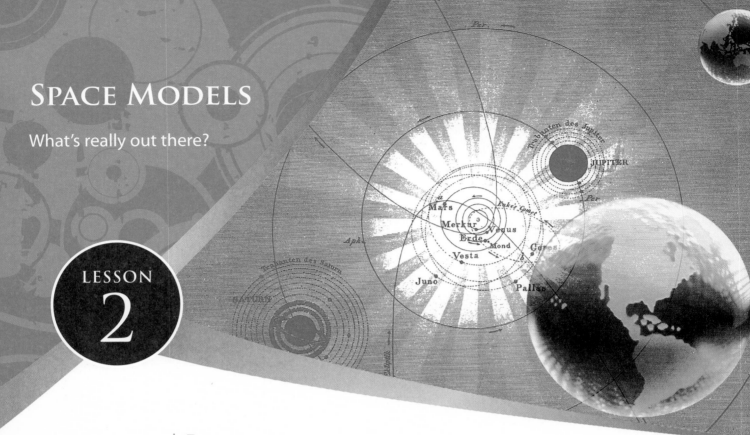

**How do we know
what our solar
system looks like?**

Words to know:

geocentric model

heliocentric model

law of gravitation

gravity

BEGINNERS

Is the earth moving? If you look outside your window at the ground, you might think the answer is no, but actually, everything in the universe is moving. You can see this from the fact that the sun rises in the east each morning and sets in the west each evening. You can see that the moon moves through the sky each night, too. If you went outside and watched the stars for a long time, you would also see that the stars appear to move through the sky as well. Early scientists noticed these movements and decided that everything in the universe must move around the earth.

However, as scientists did more testing and made more observations, they discovered that the earth and other planets are actually moving around the sun. They realized the moon moves around the earth, and that the moon, sun, and planets are all moving around in the universe.

The force that keeps all of the planets moving around the sun and keeps everything in its proper place is called gravity. You cannot see gravity, but you can see its effects. If you drop something it will fall to the ground, and you don't float away from the earth because gravity is keeping things in place on the earth, too. God created gravity to keep everything in its place.

- Does the sun move around the earth or does the earth move around the sun?

- What force keeps all the planets, moons, and stars in their places?

Have you ever played with a model car or a model airplane? Have you ever built a model train or seen a model space ship? A model is a smaller version of the real thing. It allows you to see and touch something that is too big to actually hold or play with. Space is much too big to hold or even to see completely. So man has invented models to help us understand what the universe looks like and how it works. A model of space can be very useful. These models are often drawings and not three-dimensional toys, although we sometimes see three-dimensional models of our solar system. But how did people figure out what the universe and our solar system looked like and how they work?

Geocentric model

One ancient model of the solar system was based on what is called the geocentric model, which means that the earth was believed to be the center of the universe. This model was developed based on several observations. First, the earth appears to be stationary while the sun, moon, and stars seem to move around it. The sun rises in the east and sets in the west. The moon also rises in the east and sets in the west. And the stars move across the sky. Therefore, it made sense to early observers that the earth was in the center and everything else moved around it.

Careful observation also revealed that the sun, moon and the five visible planets appeared to move among the stars. So the early model showed the earth in the center with the sun, moon, and planets each in its own sphere spinning around the earth. The stars were believed to be in the outermost sphere. The spheres were thought to be crystal or some other transparent material that allowed the people on earth to view the

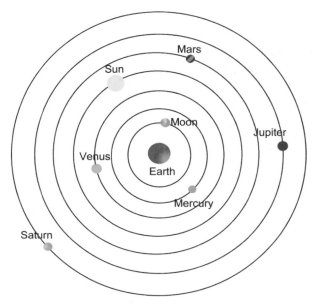

Geocentric model

objects in them.

This model was developed over several hundred years. A Greek scientist named Ptolemy did much of the work, and the geocentric model is sometimes referred to as the Ptolemaic model. However, Ptolemy and others made observations that did not fit well within the theory. Sometimes planets seemed brighter and nearer, and at other times they seemed dimmer and farther away. Also, the planets

The motion of stars across the sky can be seen with a long exposure photograph.

LESSON 2 OUR UNIVERSE · 11

Models & Tools

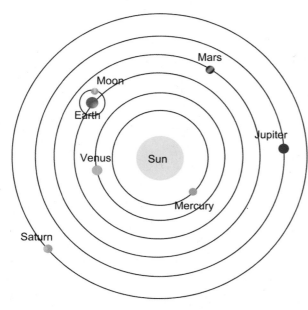

Heliocentric model

sometimes appeared to slow down and even move backward with respect to the stars. To accommodate these observations, Ptolemy shifted the earth so it was not in the exact center of the model. He then said that the planets moved in small circles within their spheres to account for the apparent backward motion. This model did not fully explain all of the inconsistencies that were observed, but it was the best model available and, for centuries, was accepted as the way the universe was.

Heliocentric model

Then, during the Renaissance, there was a renewed interest in art, science, and learning. Many scientists began to make careful observations of the heavenly bodies, and new ideas began to emerge. A Polish astronomer named Nicolaus

GRAVITATIONAL PULL

Purpose: To demonstrate that the pull of gravity is not dependent on the weight of the object.

Materials: ping-pong ball, golf ball, piece of paper, book

Procedure:

1. Hold a ping-pong ball and a golf ball. Which one is lighter?

2. Hold them both at the same height and release them at the same time. Which one hit the ground first?

3. Hold a book and a small, flat sheet of paper out at the same height and release them at the same time. Which one hit the floor first? Why?

4. Place the paper on top of the book and drop them at the same time. Did the paper float down slowly this time?

5. Finally, crumple the piece of paper into a small ball. Now, hold the book and the paper ball at the same height and drop them at the same time. Did they land at the same time?

Questions:

- Why did the book hit the ground before the sheet of paper?

- Why did the sheet of paper on top of the book stay with the book?

- Why did the crumpled piece of paper hit the ground at the same time as the book?

Conclusion:

People used to think that heavier items fell faster than lighter items—that seems logical, right? Our activity showed this to be false. The sheet of paper floated down because of air resistance, not the pull of gravity. The paper ball did not have as much resistance to the air as the flat sheet of paper did, so it fell at about the same rate as the book. Understanding gravity is important because the force of gravity pulls objects down, holds our atmosphere in place, and keeps planets in orbit around the sun.

Johannes Kepler

Copernicus developed the idea that the sun, not the earth, was the center of the solar system. This has been called the Copernican or heliocentric model. His model was able to explain many of the problems that had been observed in the geocentric model. The earth moving around the sun just like the other planets would explain why sometimes the planets appeared to move backward. The earth would catch up with slower moving planets, causing them to seem to slow down. Then after passing them, the planets' forward motion could be seen again.

Other scientists that followed Copernicus built on his foundation and were able to explain even more of what was observed. Johannes Kepler was a mathematician who very carefully plotted the movements of the planets and proved that the planets move in elliptical (or stretched) orbits instead of circular orbits, which helped explain why planets sometimes appeared closer than at other times.

The same year that Kepler published his work, another scientist, Galileo Galilei, designed and built his first telescope. He was the first one to study the heavens with a telescope. This invention allowed for much more precise measurements of the heavenly bodies and even better understanding of the workings of the planets and stars.

Galileo Galilei

Finally, in the late 1600s Sir Isaac Newton used his knowledge to explain how all of these heavenly bodies were able to move the way they do. He devised his law of gravitation, which explained how gravity helps to hold all of the planets in their orbits around the sun. Throughout the years, many improvements have been made to these theories, but the basic ideas of the Copernican model have remained, and today we have a model that explains many of the workings of our solar system.

Law of gravitation

Newton's law of gravitation states that everything exerts a pull on everything else. The more massive something is, the stronger its gravitational pull, and the closer something is to an object, the stronger its gravitational pull. Because the earth is very large and very close to us, it has a strong gravitational pull on everything on the surface of the earth.

The earth and moon exert a gravitational pull on each other. Because the earth is much larger than the moon, the moon orbits the earth. Similarly, the sun and earth exert a gravitational pull on each other. But because the sun is much more massive than the earth, the earth revolves around the sun.

Although Newton is credited with proving the law of gravity, he was not the first to recognize that the earth pulls down on objects, nor was he the first to do experiments to test the pull of the earth. Galileo did many experiments by climbing to the top of the Tower of Pisa (shown at left) and dropping various objects over the side. As he measured how long it took for them to reach the ground, he found that the pull of gravity was the same regardless of the weight of the object. ■

WHAT DID WE LEARN?

- What are the two major models that have been used to describe the arrangement of the universe?
- What was the main idea of the geocentric model?
- What is the main idea of the heliocentric model?
- What force holds all of the planets in orbit around the sun?

TAKING IT FURTHER

- Which exerts the most gravitational pull, the earth or the sun?
- If the sun has a stronger gravitational pull, then why aren't objects pulled off of the earth toward the sun?

RESEARCH THE SCIENTISTS

Choose one of the following scientists to research.
- Hipparchus
- Ptolemy
- Tycho Brahe
- Johannes Kepler
- Galileo
- Sir Isaac Newton
- Nicolaus Copernicus

Try to find the answers to the following questions:

1. When did he live?

2. What was the accepted space model at that time?

3. What problems were observed with the accepted model?

4. What contributions did he make to the space model that

he believed in?

5. What arguments did he have to answer and how did he answer them in support of the model he believed in?

Write up your answers and present them to others so they can have a better understanding of how we developed the model of space we have today.

NICOLAUS COPERNICUS

1473–1543

Nicolaus Copernicus is known as the person who changed the way the world views the universe. However he was not always known as "Nicolaus Copernicus," which was the Latin form of his name. His birth name was Mikolaj Kopernik or Nicolaus Koppernigk.

Nicolaus was born in 1473, in Poland. His father traded in copper and was a magistrate. When Nicolaus was about 10 years old, his father died and his uncle Lucas Waczenrode took him and his family in. His uncle was a canon, or clergyman, at the time.

When Nicolaus was about 15, his uncle sent him to a Cathedral school for three years. After that, he and his brother went on to the University of Krakow. Nicolaus studied Latin, mathematics, astronomy, geography, and philosophy. He later said that the university was a big factor in everything he went on to do. It was there that he started using his Latin name.

After four years of study, he returned home without a degree, a common practice at the time. His uncle wanted him to have a career in the church. To give him the needed background, he sent Nicolaus to the University of Bologna (Italy) to get a degree in canon law. While there, through the influence of his uncle, he was appointed canon at Frauenburg, which provided him with a nice income.

Shortly after this, he asked his uncle if he could return to school to complete his law

degree and to study medicine. His uncle probably would not have let him go if he were not going to study medicine. The leader of the Cathedral Chapter thought it worthwhile and gave him the necessary funds. The study of medicine may have been an excuse to study his real passion, astronomy. At this time astronomy was little more than astrology and, therefore, used in medicine. It is not known if Nicolaus ever completed his medical training, but upon returning home, he worked for about five years as his uncle's doctor.

Although he worked as a doctor, Copernicus continued to study astronomy. At the time, most people believed that the earth was the center of the universe and all heavenly bodies orbited around it. However, Copernicus came to a different conclusion based on his studies of the heavens. And in 1514 he distributed a handwritten book on astronomy, without an author's name in it. The book made the following points:

1. The earth's center is not the center of the universe; the center of the universe is near the sun.

2. The distance from the earth to the sun is imperceptible compared with the distance from the earth to the stars.

3. The rotation of the earth accounts for the apparent daily rotation of the stars.

4. The apparent annual cycle of movements of the sun is caused by the earth revolving around it.

5. The apparent retrograde motion of the planets is caused by the motion of the earth from which one observes the planets.

These ideas were revolutionary and not commonly accepted, which is why Copernicus published his book without his name in it. This little book was a precursor to his major work, which was completed at the end of his life. Copernicus did not spend his whole life studying and writing, however. In 1516, Copernicus was given the administrative duties of the districts of Allenstein for four years. In 1519, when war broke out between Poland and the Teutonic Knights, he was in charge of defending his area. In 1521 he was able to successfully lead the defense of Allenstein Castle, and an uneasy peace was restored. He was next appointed Commissar of Ermland and given the job of rebuilding the area after the war. Around 1522, he returned to Frauenburg and finally got the peaceful life he was looking for.

Even during the war, Copernicus continued his observations of the heavens. And after returning to Frauenburg, he began to work continuously on his book. Copernicus's theory of the solar system may have remained unknown, however, if not for a young Protestant named Rheticus who came to visit Copernicus in 1539. It was an unusual thing for a Protestant to visit a Catholic stronghold at this time, but Rheticus lived with Copernicus for about two years and helped him get his book published.

Rheticus took the manuscript to a printer named Johann Petreius in Nürnberg. He was unable to stay around and watch over the printing of the book, so he asked a friend named Andreas Osiander, a Lutheran theologian, to oversee it. Andreas Osiander removed the introduction letter originally written by Copernicus, and inserted his own. This substitute letter said that the book was to be used as a simpler way to calculate the positions of the heavenly bodies and not to be taken as truth. Copernicus received his first copy of the book on the day he died, so the switch was not discovered for 50 years.

When Osiander's switch was discovered, some people were appalled; others felt it was the only reason the work was not immediately condemned by the Catholic Church. Regardless of the reasons for the switch, the publication of the book changed the way man looks at the universe. Copernicus's work went on to inspire Galileo and Newton and generations of scientists to follow.

THE EARTH'S MOVEMENT

Rotating and revolving

LESSON 3

How does the earth move and how does that affect us?

Words to know:

rotation

revolution

solstice

equinox

Challenge words:

Foucault pendulum

BEGINNERS

In the last lesson you learned that the earth is moving, even though it doesn't feel like it. The reason it doesn't feel like the earth is moving is that you are moving with the earth, at the same speed that the earth is moving.

The earth is moving in two different ways. First, the earth is spinning like a top. If you put a toothpick through a ball of clay and then spin it, the clay will spin around the toothpick. The earth does not have a stick or pole through the middle of it, but it is spinning just like the ball of clay. We call this spinning motion **rotation**. The rotation of the earth is what makes the sun appear to rise, move through the sky, and then set. The earth rotates or spins all the way around one time each day, so the sun shines on different parts of the earth as it rotates.

The earth is also moving around the sun. This is called orbiting or **revolving** around the sun. It takes the earth one year to orbit the sun. The earth is tilted in space. During the summer, the area of the earth where you live is tilted toward the sun, so the sun shines more directly there; thus it is warmer outside. During the winter, your part of the earth is tilted away from the sun, so the sunshine comes to the earth at more of an angle; thus it is colder outside. The earth's tilt and movement around the sun cause us to experience the seasons.

- In what two ways is the earth moving?

- Why do we experience seasons like summer and winter?

Various models of the universe were based upon observations about how the earth and other heavenly bodies move. When the geocentric model was the accepted model, scientists observed that the sun rose in the east and set in the west and that a day was 24 hours long. They also observed that the stars appeared to move through the sky. This led them to believe that these objects moved around the earth. However, these observations can be accounted for in the heliocentric model as well by showing how the earth moves with respect to the sun and the stars.

The earth's movements

Once the heliocentric model was accepted, scientists began to make very careful measurements of how the earth moves. The earth moves in two different ways. First, it rotates on its axis, an imaginary line going from the North Pole to the South Pole through the center of the earth. Second, the earth revolves around the sun.

Rotation

The rotation of the earth on its axis explains how we experience the rising and setting of the sun and the relative movement of the stars. It also explains other observations that could not be easily explained by a stationary earth.

The earth bulges slightly around the equator. The diameter of the earth measured at the poles is approximately 7,900 miles (12,713.56 km), but the diameter at the equator is about 7,927 miles (12,756.28) km. This difference is caused by rotation. It is said to be caused by centrifugal force—a force generated by spinning objects that causes them to pull outward. Accurate measurements of Jupiter and Saturn show that they bulge even more around the center because they are more massive and spin faster than earth. On the other hand, Mercury and Venus bulge less around the center

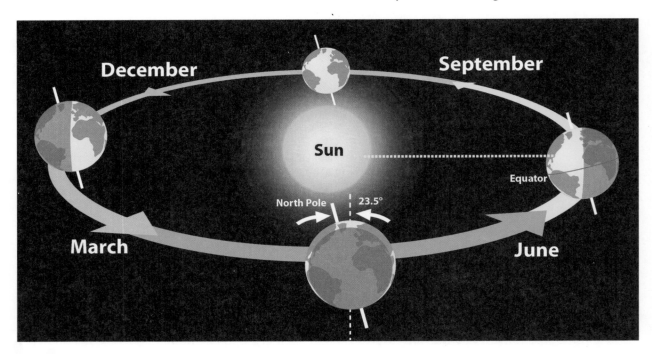

DEMONSTRATING MOVEMENT

Activity 1—Purpose: To demonstrate how a rotating earth gives hours of daylight and darkness

Materials: masking tape, volleyball or basketball, flashlight

Procedure:

1. Place a small piece of masking tape on a basketball or volleyball then hold the ball out in front of you.

2. Have another person hold a flashlight representing the sun and shine it on the ball. Slowly rotate the ball.

3. Observe when the light is shining on the piece of tape and when the tape is in the shadow or darkness. This shows how we experience day and night.

Activity 2—Purpose: To demonstrate the observed parallax of stars

Materials: None

Procedure:

1. Hold your arm straight out in front of you with one finger pointing up.

2. Using only your right eye, look at a distant object, noting where your finger is with respect to the object.

3. Close your right eye and look at the object with your left eye. Note how your finger appears to shift with respect to the distant object.

4. Repeat several times, alternating which eye is open.

Conclusion:

The different locations of your eyes represent the different locations of the earth with respect to the stars. It is easy to see the finger appear to move because the finger and the object are both relatively close to you. However, this effect is harder to see with the stars because they are so far away from the earth.

because they spin at a slower rate than the earth.

Another indication of a spinning planet is the movement of air and water masses on earth. The rotation of the earth causes something called the Coriolis effect. Hot air near the equator rises and colder air from the poles moves in to take its place. Without the rotation of the earth, this air would move in vertical lines from the equator to the poles and back. However, what is observed is a diagonal airflow with respect to the earth's axis. This is caused because the air near the equator moves faster than the air near the poles due to the rotation of the earth. The Coriolis effect due to the rotation of the earth also causes cyclones to spin counter-clockwise in the northern hemisphere and clockwise in the southern hemisphere.

Revolution

The second way the earth moves is by revolving around the sun. The path the earth takes in its revolution around the sun is called its orbit. When scientists accepted the geocentric model, they accounted for changing seasons by saying that the sun moved differently around the earth at different times of the year. But seasons are better explained by the earth orbiting the sun and the tilt of the earth's axis. The earth is not completely vertical with respect to the sun. The earth is tilted 23½ degrees from vertical. When the northern hemisphere is tilted toward the sun, the sun's rays are more direct, causing warmer temperatures in that hemisphere during the summer. When the northern hemisphere is tilted away from the sun, the sun's rays hit at a steeper angle, causing less heating and therefore lower temperatures in the winter.

The four parts of the earth's orbit

The earth's orbit is divided into four parts. The summer solstice, the first day of summer, occurs on June 21st, when the sun's rays directly hit the Tropic of Cancer, which is an imaginary line around the earth at 23½ degrees north of the equator. The earth reaches the halfway point in its orbit on December 21st, the first day of winter, or the winter solstice. This is when the sun's rays directly hit the Tropic of Capricorn at 23½ degrees south of the equator. Halfway between these points is the spring equinox, the first day of spring, which occurs on March 21st, and the autumnal equinox, the first day of autumn, which occurs on Sept. 22nd.

Other observations also point to a moving earth. First, with powerful telescopes, scientists have observed something called parallax of the stars. This is where stars that are closer seem to shift their positions with respect to stars that are farther away as the earth moves through space. In fact, a lack of observed parallax was one argument against Copernicus when he first suggested the heliocentric model. He argued that there was a parallax but that they could not observe it because the distances to the stars were too great. He has been proven correct with the invention of powerful telescopes that can now measure these apparent movements.

Finally, it has been observed that more meteors and brighter meteors are observed after midnight than before midnight. This occurs because as the earth rotates on its axis, the area on the earth where it is between midnight and sunrise is on the forward part of the orbit. It is moving toward oncoming debris in space and more collisions will be observed during this time. These observations, as well as many others, support the heliocentric model and demonstrate the rotation and revolution of the earth. ■

A meteor streaks across the sky.

WHAT DID WE LEARN?

- What are the two different types of motion that the earth experiences?
- What observations can we make that are the result of the rotation of the earth on its axis?
- What observations can we make that are the result of the revolution of the earth around the sun?
- What is a solstice?
- What is an equinox?

TAKING IT FURTHER

- What are the advantages of the earth being tilted on its axis as it revolves around the sun?
- One argument against Copernicus's theory was that if the earth were moving, flying birds would be left behind. Why don't the birds get left behind as the earth moves through space?

FOUCAULT PENDULUM

Even after scientists concluded that the earth must rotate on its axis, it was very difficult to demonstrate this movement. One idea was to drop a rock down a very deep shaft and see if it moved sideways compared to the shaft. This did not work because the depth of the shaft was very small compared to the radius of the earth, so the sideways movement was too small to measure. A similar experiment was to fire a cannon ball north to south and measure its movement east to west. Again however, the movement was too

FUN FACT

In 1852, Leon Foucault also invented the gyroscope, a special kind of top that is used in many aerospace applications.

small to measure.

Eventually however, a French scientist

named Leon Foucault devised a way to demonstrate the rotation of the earth. Foucault used a very long pendulum which would swing slowly. He placed marks in a circle on the floor below the pendulum. This allowed an observer to watch the path of the swinging pendulum. Over time, the pendulum appeared to change its path, but what was actually happening was the earth was rotating under the pendulum, thus moving the marks in the circle. This device is called a **Foucault pendulum**.

Purpose: To demonstrate the movement of the earth with your own Foucault pendulum

Materials: copy of clock pattern, needle, thread, tape, modeling clay, tripod, swivel chair, stool, or turntable

Procedure:

1. Tape a copy of the clock pattern to the top of a swiveling chair, stool, or other turntable.

2. Place a tripod on top of the chair and center the tripod over the circle.

3. Cut a length of sewing thread long enough to reach from the center of the tripod to the clock circle.

4. Thread one end of the string through a needle and tie a knot in the thread to prevent the needle from slipping off.

5. Push the needle through a ½-inch ball of modeling clay so that a small amount of the

needle sticks through the end of the clay.

6. Tape the other end of the thread to the center of the tripod so that the needle swings freely just above the circle.

7. Start the thread gently swinging across the circle from 12 to 24.

8. Smoothly turn the chair ¼ of a turn to the right. Observe the numbers that the needle is now swinging across.

9. Turn the chair ¼ of a turn more and

observe the numbers over which the needle is swinging.

10. Repeat until a complete turn has been made.

Questions:

• What forces are affecting the pendulum?

• Why does the pendulum eventually stop moving?

• How does a Foucault pendulum keep moving for hours or days at a time without stopping?

Conclusion:

Although the tripod and circle are moving with the chair, the swinging of the thread is mostly unaffected by its turning. Full-size Foucault pendulums can be seen in many museums and they demonstrate the rotation of the earth in much the same way as your smaller model.

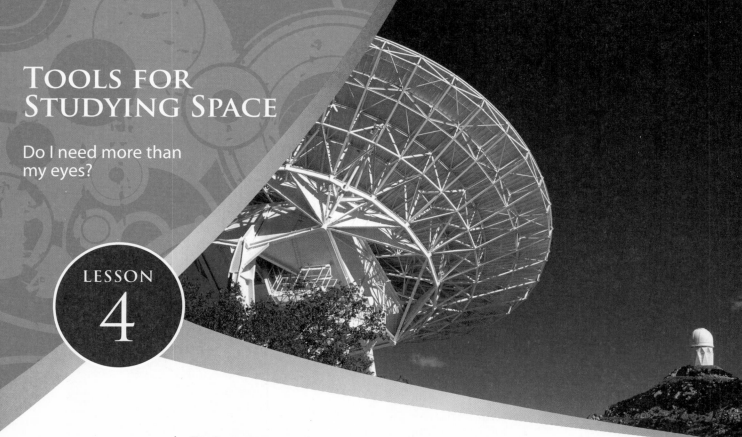

TOOLS FOR STUDYING SPACE

Do I need more than my eyes?

LESSON 4

What tools are used to study the universe?

Words to know:

refracting telescope

reflecting telescope

radio telescope

Challenge words:

interferometry

BEGINNERS

Do you like to sit outside at night and look at the stars? People have always enjoyed watching the stars. The Bible tells us that "the heavens declare the glory of God; and the firmament shows His handiwork" (Psalm 19:1). God created the beautiful night sky, and we can enjoy learning about the stars. We can see many stars just by looking up at the night sky, but there is more in space than just the stars that you can see with a naked eye.

Scientists have developed many scientific instruments to help them better observe the universe. One of the most important instruments for looking at space is the telescope. A telescope uses lenses and/or mirrors to make something look bigger. This allows us to see things in space in much greater detail. With a telescope, you can see the surface of the moon or some of the planets in our solar system. Telescopes also help scientists to see stars that are too far away or too dim to see with just the eye. This allows scientists to better understand how the stars and planets move through space.

• **What is a telescope and what is it used for?**

From the very beginning, man has enjoyed gazing at and studying the stars. Yet there is a limit to what man can understand about the universe using only his eyes. Many instruments have been developed to help track the movement of the sun, stars, and planets, as well as to view and measure the distant parts of the universe.

One of the oldest existing structures speaks of man's attempt to understand the universe. Stonehenge (shown here) is an area in England where large stone slabs have been raised. The builders of Stonehenge are unknown and the exact nature of the use of Stonehenge is also unknown. However, the stones line up in such a way as to mark the summer solstice. These stones track the movement of the earth around the sun. Many other ancient civilizations also built buildings or monuments that mark the seasons.

Stonehenge

Ancient devices

Another ancient attempt to track the movement of the sun is the sundial. This device used the shadow cast by the sun as it moved through the sky to indicate the time of day. The changing angle of the shadow from day to day helped to indicate the seasons. Sundials were used for centuries, until the invention of more accurate clocks.

Sailors used a device called a quadrant (shown here) to mark their location with respect to the stars. A quadrant is a device shaped like one fourth of a circle, with an angular scale and a moveable sight. The quadrant was used to measure the angle that a particular star made with the horizon. Coupled with accurate charts of the stars, a sailor could measure his latitude (distance north or south) with a quadrant. This navigation device was extremely useful and necessary on sea voyages that left sight of land.

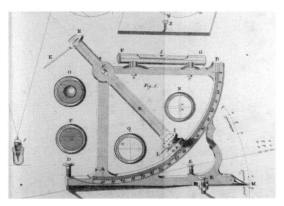

A diagram of a quadrant

Refracting telescopes

But man wanted to know more about the universe. In 1608 Hans Lippershey, a Dutchman, invented the first practical telescope. And in 1609 an Italian scientist named Galileo Galilei built his own telescope and became the first man to use a telescope for viewing the heavens. His was a very simple telescope with a lens at either end of a tube. These lenses magnified the image, allowing more accurate viewing of the stars and planets. The invention of the telescope started a new age of space exploration.

Galileo's telescope was called a refracting telescope. Refracting telescopes use two lenses. The first lens refracts or bends the light from the distant star to make a concentrated image. The second lens, the eyepiece, bends the light once more to make the beams parallel

1. Refractor

again. One problem encountered with this type of telescope was that false colors, called chromatic aberrations, appeared around the image due to the bending of the light. To avoid this problem, the lenses were made thinner so they did not bend the light as much, but this required the telescopes to be much longer. Some telescopes were as long as 200 feet. Today, refracting telescopes use a series of lenses that focus the different colors of light at the same point, thus reducing the false colors.

Reflecting telescopes

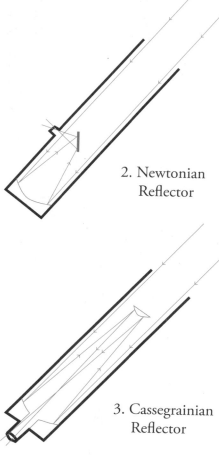

2. Newtonian Reflector

3. Cassegrainian Reflector

Sir Isaac Newton discovered the reason for this chromatic aberration, or false color. It was caused because white light is composed of all colors of light combined, and as the lens bends the light, the different colors bend at different angles. So Newton decided to build a telescope that avoided this problem. He invented the reflecting telescope, which uses a combination of mirrors and lenses. Instead of a lens to collect the light, a Newtonian reflecting telescope uses a concave mirror to collect the light and a flat mirror to project the image on the side of the telescope where the eyepiece is.

Newton's design requires that the image be viewed from the side of the tube. A different arrangement of mirrors, called the Cassegrainian reflector, uses a concave mirror at the end of the tube to collect the light, then a convex mirror in the middle to reflect the light back to the end where the eyepiece magnifies the image. This arrangement allows the image to be viewed from the end of the telescope.

> ## FUN FACT
>
> The mirror on the Large Binocular Telescope located on Mt. Graham in Arizona weighs over 16 tons.

Modern telescopes

The basic design of the early refracting and reflecting telescopes is still used today. However, many improvements have been made to the designs of the mirrors and lenses, and computer technology has allowed scientists to accurately control these telescopes and to film the far reaches of space.

A much more recent invention is the non-optical telescope, or radio telescope. These devices, such as the ones shown here, detect radio waves that are emitted from distant stars. They collect and concentrate these waves so scientists can view other characteristics of stars that cannot be viewed with the eye. These telescopes can also send radio waves and detect what is returned, much like sonar or radar, and collect more information about neighboring planets.

Space telescopes

One of the most important advancements in telescope technology has been the launching of the Hubble Space Telescope in 1990, shown here as it was being released from the space shuttle cargo bay. This telescope was placed in orbit around the earth to allow astronomers to view the stars without the interference of the earth's atmosphere.

The Hubble Space Telescope (shown here) has an 8-foot wide mirror and was built at a cost of $1.6 billion. The images from the Hubble are sent to the Space Telescope Science Institute in Baltimore, Maryland, where they are analyzed by astronomers from around the world. In addition to the mirror,

REFRACTION & REFLECTION

Purpose: To demonstrate refraction and reflection

Materials: flashlight, magnifying glass, mirror

Procedure:

1. Go into a dark room

2. Shine the beam of a flashlight through a magnifying glass onto a wall.

3. Observe how the beam is concentrated. Look for rainbows around the edges of the beam's outer circle.

4. Next, shine the beam of the flashlight on a mirror at an angle so that it shines onto a wall.

5. Observe the angle at which the beam reflects from the mirror. What happens to the color of the beam?

Conclusion:

The magnifying glass is a convex lens just like the refracting lens of a refracting telescope. The white light from the flashlight is separated around the edges, and shows up as rainbows on the wall. This is similar to the chromatic aberration that was observed with refracting telescopes.

When reflected with the mirror, the light should remain white. You should not see any rainbows in the reflected light. Most household mirrors are flat mirrors, which just reflect the light at the same angle at which it entered the mirror. This is the kind of mirror that reflects the image to the eyepiece in a Newtonian telescope. Concave mirrors, ones that are curved in toward the center, actually concentrate the light and reflect it in a stronger beam. A concave mirror is the type located at the end of a reflecting telescope.

Optional Activity:

If you have access to a telescope, you can use it to view objects around you during the day, and to view the stars and the moon at night (**but never look directly at the sun**). Some of the most interesting things to see with a telescope are the craters on the moon. Find out what you can about the design of the telescope. Is it a refracting or reflecting telescope? Do you view the objects from the side or the end of the telescope? How powerful are the lenses—how much do they magnify the image?

Hubble is fitted with instruments that determine the temperature and chemical make-up of distant objects, as well as instruments that measure ultraviolet light, brightness and infrared light.

When the Hubble Telescope was first launched from the space shuttle *Discovery*, it did not work properly. It was discovered that even after a year of polishing, the telescope's mirror had a flaw that prevented it from focusing the light properly. So, in 1993, another shuttle mission was conducted to repair the telescope. This mission was a success and the Hubble immediately began sending back astounding images. A third shuttle mission was conducted in 1997 to upgrade the Hubble's systems. This telescope has provided, and continues to provide, startling new pictures of objects in space and huge amounts of data for astronomers to analyze.

A new space telescope is currently being designed. The James Webb Space Telescope (JWST) is scheduled to be launched in 2013 with the purpose of getting information about stars that are farthest away from us. It will operate primarily in the infrared light spectrum. Its mirror will have a diameter that is 2.5 times bigger than the Hubble and will be able to detect objects that are 400 times fainter than any technology currently available can detect. JWST will have a 5–10 year life span. JWST is a joint project conducted by the United States, Canada, and the European Space Agency. ■

WHAT DID WE LEARN?

- What are the three main types of telescopes?

- What was one disadvantage of the early refracting telescope?

- How did Newton avoid this problem?

TAKING IT FURTHER

- Why do you think scientists wanted to put a telescope in space?

- What kinds of things can we learn from using optical telescopes?

- What kinds of things can we learn from radio telescopes?

TELESCOPE ADVANCES

Since the invention of the telescope, scientists have understood that the larger you make the opening to the telescope, the more light it will gather, and thus the more you can magnify the image with less distortion. However, there are limits to how big you can make a mirror or lens for a telescope. The bigger the mirror is, the more it becomes distorted by gravity pulling on its own mass. Thus scientists have developed ways to make mirrors larger without increasing this distortion.

The two giant telescopes used at the Keck Observatory on top of Mauna Kea, a mountain on the big island of Hawaii, are 30 feet (10 m) across (pictured at left). A single mirror of this size would have significant distortion; however, these mirrors are actually a series of hexagon-shaped mirrors (shown here) that are precisely fitted together to form one large mirror. Each segment is supported by a structure that can adjust its position with respect to the segments around it. These adjustments are made twice each second. This results in a very large mirror that can gather light from very distant stars and allow the scientists to see more clearly into space.

The large mirror design is only part of the wonder of the Keck telescopes, however. The two telescopes sit 275 feet (85 m) apart. The light from these two telescopes can be used together to give the equivalent images that a 275-foot telescope could give. The process of combining the light from the two telescopes is called **interferometry**. The resulting image is brighter and sharper than the image from either of the telescopes individually.

FUN FACT

Mauna Kea is home to more than 13 scientific telescopes. Most are optical telescopes, but there are infrared and radio telescopes there as well.

Purpose: To demonstrate how light pollution can interfere with telescopes

Materials: sheet of paper, black marker, car

Procedure:

1. On a sheet of white paper, use a black marker to write letters like an eye doctor's chart with large letters at the top and smaller letters at the bottom.

2. At night, have someone stand in front of a parked car that has its lights off and hold the sign up.

3. Stand several feet away and shine a flashlight at the sign. Read as many of the letters as you can.

4. Next, have someone turn on the car's headlights and have the person holding the sign, hold it just above one of the headlights.

5. Again, shine your flashlight at the sign and see how many of the letters you can read.

Question:

When was it easier to read the letters, with the headlights on or off? Why?

Conclusion:

The light from the car interferes with your ability to see the sign. This is called light pollution and is why telescopes are located away from the bright lights of the city and are often on the tops of mountains in areas with few people. Light in the area around the telescope can interfere with viewing the stars.

GALILEO GALILEI

1564–1642

Have you ever gone camping and looked at the sky on a clear night? It can take your breath away. The number of stars you see are past counting. The beauty and splendor of what God has put there for us to enjoy has been the topic of stories and wonder ever since God made the world.

Many men and women have looked at the stars and wanted to know more. Maybe it's because this makes them feel closer to God. Whatever the reason, Galileo Galilei was no exception. He looked at the stars and wanted to know more. What made Galileo exceptional is what he did to learn more about the stars.

Galileo was the son of a professional musician who liked to experiment on strings. Galileo was born in 1564, when his father was 44 years old. His parents felt that Galileo's mathematical and mechanical pursuits did not promise a substantial return. They wanted him to follow a more suitable profession—one in the medical field. But their hope was in vain.

Galileo did not follow the prevailing system of learning, which was always to learn by reading what the authority said on a subject and accept it as truth. To learn about nature, he was told to read Aristotle and accept what he had written as the final authority. However, Galileo wanted to learn through experiments and calculations.

This way of doing things brought much controversy to his life.

In 1609 Galileo heard about a spyglass that was invented by a Dutchman and was being demonstrated in Venice. From the reports about this spyglass, and his own understanding of mathematics, he was able to build a telescope with which to observe the heavens. He is believed to be the first to use this technology to observe the stars. The first telescope he built only magnified objects 3 times, making them appear three times bigger than they appear to the naked eye. After more work, he was able to magnify objects 32 times. He later modified the telescope to view very small things, using it as a microscope, or "occhialini" as he called it. He made several of these and gave them to various people to use.

With his telescope, Galileo was able to see many new things in space that had not been seen before. He discovered the moons, or satellites, of Jupiter and saw their orbit around the large planet. Using the telescope to study Venus led to his understanding of how our solar system works, and led him to accept the model of the solar system developed by Nicolaus Copernicus. As Galileo studied the heavens, he noticed that Venus had phases like our moon. This led him to believe that the planets went around the sun and not around the earth. This was called the Copernican model and it was not well received by the leading scientists of the day. They had been influenced by Greek philosophy and viewed Galileo as endangering their positions of authority.

Many in the Church believed that the Bible taught that the earth is the center of the universe, and so felt that Galileo's theories would undermine the authority of the Bible. They had interpreted the Bible to agree with the secular Ptolemaic view of the solar system. Others in the Church were at first very open to Galileo's theories, but later, due to political and personal reasons, Pope Urban VIII issued a decree that Galileo be tried for heresy (false teaching).

In 1616 he was tried by the Catholic Church and found guilty of heresy. The Church put him in prison until he died in 1642. But this was not the type of prison you might think of today. He spent most of this time under house arrest, living in the homes of his friends. Throughout this ordeal, Galileo held a strong belief in God and had a deep respect for the Church. He wrote, "I have two sources of perpetual comfort—first, that in my writings there cannot be found the faintest shadow of irreverence towards the Holy Church; and second, the testimony of my own conscience, which only I and God in heaven thoroughly know. And He knows that in this cause in which I suffer, though many might have spoken with more learning, none, not even the ancient Fathers, have spoken with more piety or with greater zeal for the Church than I."

His arrest may have slowed him down but it didn't stop him from his work. He continued to perform experiments and by the end of his life he had given science many ideas that would prove very beneficial. One of the most important ideas discovered by Galileo was the idea of inertia. This theory states that a body in motion will remain in motion unless something acts on it to stop it. This was in direct conflict with Aristotle's view that some force must continuously act on a body to keep it in motion. Galileo was later proven right.

As smart as Galileo was, he still made mistakes. He thought that the tides were caused by the earth's rotation on its axis. He did not consider the gravitational pull of the sun and moon. He also thought that comets and meteors were atmospheric phenomena instead of heavenly bodies outside the earth's atmosphere. But in spite of his mistakes, Galileo contributed greatly to our understanding in many areas of science and is considered by many as the "father of modern physics."

(For more about the "Galileo Controversy," see www.answersingenesis.org/go/galileo.)

UNIT

2

OUTER SPACE

KEY CONCEPTS

- ◊ **Describe** the structure of the universe and galaxies.
- ◊ **Define** the relationship between stars and constellations.
- ◊ **Identify** the differences that exist among stars.
- ◊ **Compare** and **contrast** asteroids, comets, and meteors.

UNIT LESSONS

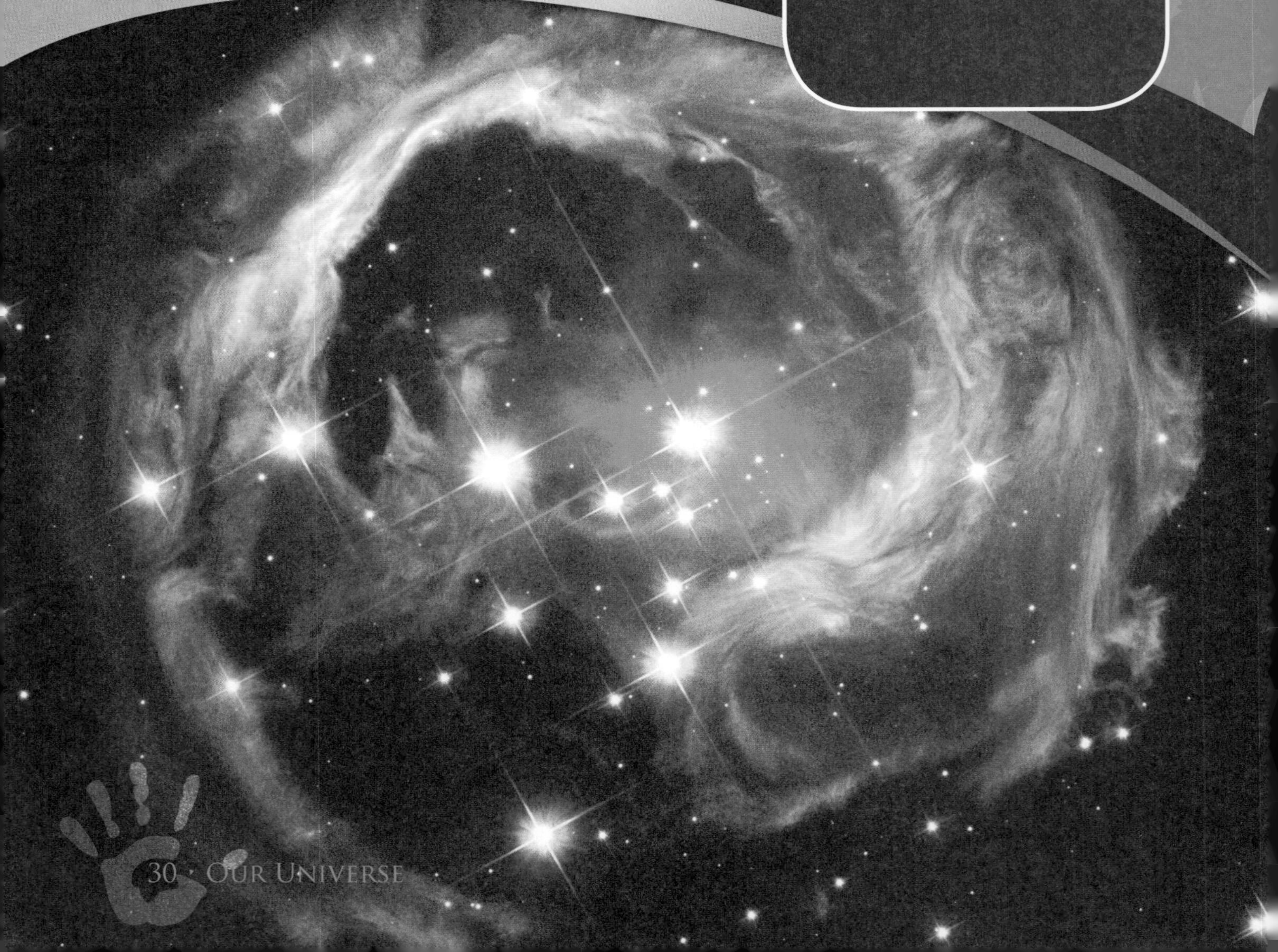

OVERVIEW OF THE UNIVERSE

How big is it?

LESSON 5

Where do we live in space and how big is the universe?

Words to know:

light-year

Milky Way

asterism

Challenge words:

celestial equator

prime hour circle

vernal equinox

degrees of declination

hours of ascension

BEGINNERS

God created many interesting things in space. The night sky is full of beautiful stars. Scientists believe that there may be too many stars in the universe to count. But one star is special to us; it is the sun. The sun is the closest star to earth and gives us light and heat. We already learned that the earth orbits, or moves, around the sun. There are seven other planets that also orbit the sun. The sun and the planets (and their moons) that orbit the sun are called our solar system. You will learn more about each of the things in our solar system in other lessons.

Outside of our solar system are billions of stars. We cannot see all of the stars in the universe without special telescopes, but we can see many of them just by looking at the night sky. Sailors have used the stars to help them find their way on the sea. And people use the stars to help them mark the seasons. In order to use the stars, people have identified pictures that can be made by connecting certain groups of stars together, sort of like a dot-to-dot puzzle. On the next page is a picture showing the Big Dipper and the Little Dipper. These are two of the easiest groups of stars to spot in the night sky. With practice and the help of a star chart, you can learn to recognize and identify some of the constellations, or groups of stars, too.

• What objects make up our solar system?

• What is a constellation?

• Why did sailors need to be able to recognize stars?

Exactly how big is the universe? Only God can answer that question. Some radio telescopes reach out to a distance of up to 15 billion light-years. A **light-year** is a measure of distance, not time. It is the distance that light travels in a year and is equal to about 6 trillion miles (9.4 trillion kilometers), so the known universe has a radius of at least 90 thousand billion billion miles (144 thousand billion billion kilometers). Scientists are not even sure if the universe has an end. With the invention of the telescope, man has learned a great deal about the universe. But there are still many unanswered questions, and much of what is believed to be true about the universe cannot really be tested or directly observed because of our fixed location on earth and the great distances to objects in space.

Objects in space

We know the most about our own solar system. Our solar system consists of a sun orbited by eight planets and various asteroids, moons, and other objects. Our solar system is part of the **Milky Way** galaxy. Our sun is one of millions of stars that are revolving about the galaxy center. The Milky Way is believed to be shaped like a flat disk with arms extending out like a giant pinwheel, similar to the spiral galaxy shown here. Other galaxies are elliptical, shaped like an elongated oval. Scientists believe that there are billions of galaxies in the universe, all with billions of stars, some with planets, moons, etc.

There are also other objects in the universe including nebulae, asteroids, comets, quasars, and black holes. Scientists are just beginning to understand many of these unusual objects. But one thing that people have understood from ancient times is the movement of the stars through the night sky relative to the earth. Stars have been mapped since ancient times. Ptolemy mapped 48 different constellations, or groups of stars, around AD 150. The Greeks, Romans, and Babylonians all gave names to many of the constellations in the night sky. Many of these names were based on the pictures that were supposedly formed by connecting certain stars together. Most constellations were named after mythological characters.

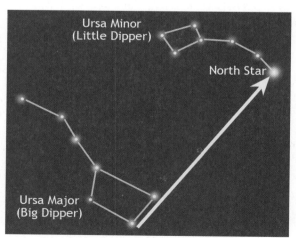

Constellations

Today, we have star charts that can help us identify many of these same constellations. The constellation Ursa Major ("Great Bear") contains the group of stars

FUN FACT
Polaris, or the North Star, which is the end of the handle of the Little Dipper, does not move with respect to the earth. This phenomenon has helped sailors in the northern hemisphere navigate on the ocean for generations.

OBSERVING THE NIGHT SKY

Review many of the most easily identifiable constellations using a good star chart. Then, on a clear night, practice using the chart to help locate and identify these constellations in the sky.

commonly called the Big Dipper. The handle of the Dipper is the Great Bear's tail and the Dipper's cup is the Bear's flank. The Big Dipper is not a constellation itself, but an asterism, which is a distinctive group of stars. Another famous asterism is the Little Dipper in the constellation Ursa Minor ("Little Bear"). Other easily identifiable constellations include Cassiopeia and Orion. By studying star charts, you can learn to locate and identify these and many other constellations.

Because of the rotation of the earth and its movement around the sun, the location of various constellations with respect to a particular spot on the earth moves throughout the night and from one day to another. A good star guide will help you by showing the location of constellations at various times of night and at various times of the year.

Many constellations that are visible in the northern hemisphere are not visible in the southern hemisphere. Sailors used to navigate by the stars and had to learn the different constellations in each part of the world they were likely to visit. By studying the universe, we begin to appreciate the amazing wonder of God's creation. ■

WHAT DID WE LEARN?

- What is our solar system?
- Our solar system is part of which galaxy?
- How big is the universe?

TAKING IT FURTHER

- Why do you think our galaxy is called the Milky Way?
- Why do you need star charts that are different for different times of the year?
- Why do you need star charts that are different for different times of the night?

LOCATING STARS

The whole earth has been divided into areas defined by the lines of latitude and longitude. If you want to locate any spot on earth, you just need the latitude and longitude coordinates. Similarly, the night sky has been defined by lines that are projected from the earth onto the sky. The plane of the equator is projected out from the earth onto the sky to define the celestial equator. The prime meridian, the longitude line passing through the poles and Greenwich, England, is projected onto the sky and called the prime hour circle. The point in the sky where the celestial equator and the prime hour circle meet is called the **vernal equinox**.

Using these lines, the night sky map is divided into sections. The lines parallel to the celestial equator are called **degrees of declination**. Stars that are found north of the equator are given a positive number and those south of the equator are given a negative number. The lines parallel to the prime hour circle are called **hours of ascension**. The night sky is divided into 24 sections or hours, and the star is designated by its location right of the prime hour circle.

To locate a particular star on the star map, you just need to know its declination and right ascension. For example, the star Sirius is located at -16.7°, 6 hours 44 minutes. This means that the star is located 16.7 degrees below the celestial equator and 6 hours and 44 minutes right ascension from the prime hour circle. With coordinates, it is fairly easy to locate a star on a star map. But actually locating the star in the sky at any given time is not so easy because the earth is always moving with respect to the stars, and a chart that takes into account the date and time must be used.

In addition to the lines of declination and ascension, the whole night sky has been divided into 88 areas defined by constellations. To an astronomer, a constellation is an area in the sky, not the picture defined by the stars in that area. The constellations are

FUN FACT

Several constellations are mentioned in the Bible. Pleiades, the Bear and her Cubs, and Orion are mentioned in Job 9:9, Job 38:31–32, and Amos 5:8.

well defined but are not square or rectangular like the areas on a globe. Instead, they are irregular to match the patterns of the stars.

The stars within a constellation are named according to the constellation. The star is given a Greek letter designation followed by a modification of the name of the constellation. The brightest or most important star in the constellation is usually called alpha, the second brightest beta, and so

on. For example, the first star in the Big Dipper (also called Ursa Major) is thus called Alpha Ursae Majoris. The second star in the dipper would then be called Beta Ursae Majoris and so on.

There are also naming systems used to identify other heavenly objects. One system is called the Messier catalog. The objects in this catalog (nebulae, galaxies, and clusters) are each numbered and referred to by their number. So the 20th entry in the catalog would be Messier 20 or M20 for short.

Questions:

- Explain how a star map is similar to a map of the globe.

- What units are used to measure declination and ascension?

- How does an astronomer define a constellation differently than most people?

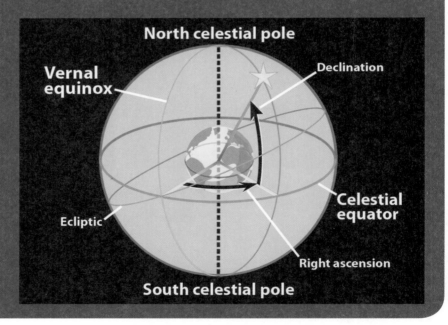

STARS

Twinkle, twinkle little star

What different kinds of stars are there?

BEGINNERS

When you look at stars in the night sky, do they all look alike? Probably not; some stars look bigger and brighter than others. Some stars look dimmer and are harder to see. Some stars are whiter than others. This is because all stars are not alike.

The closest star to the earth is the sun. By studying the sun, scientists have learned that stars are burning balls of gas. But different stars burn at different temperatures, so some are brighter than others or shine with a slightly different color.

All of the stars, except the sun, are very far away from the earth, but some stars are closer than others. Thus, some stars look brighter than others. The next time you are looking at the stars, see if you can guess which ones are closer and which ones are farther away.

• Why might some stars look different from other stars?

The early study of the stars mostly involved studying the motion of the stars through the sky and charting the constellations. But since the invention of the telescope, and with the aid of modern technology, we have been able to learn a great deal about the stars. Much of what we know about stars comes from our study of the sun. It is the closest star and therefore the easiest to study. We know that stars are hot balls of burning gas. But not all stars are the same. Looking at stars with our eyes alone reveals that some stars are brighter than others and some stars have different shades of color, while others appear to be white.

Brightness and distance

Scientists describe stars by various characteristics, including brightness, distance from the earth, color, size, and motion relative to the earth. The brightness of a star as seen from the earth depends on two things: the amount of light it emits and how far away it is. One of the first astronomers to describe the differing brightness of the stars was a Greek named Hipparchus. Hipparchus assigned a number to each star to describe its brightness. He assigned magnitudes from 1 to 6, with 1 being the brightest stars he could see and 6 being the faintest. We still use the same scale to describe brightness today. However, with the aid of telescopes we can now see stars that are much fainter than Hipparchus could see and some that are brighter, so the scale now goes from −1 (brightest) to 25 (faintest). On this scale of brightness, the full moon would be −12 and the sun would be −27.

The second characteristic of a star is its distance from the earth. The closest star to the earth, not including our sun, is called Proxima Centauri, a star that is part of the Alpha Centauri system. It is approximately 25 trillion miles away from the earth. Because stars are so far away, astronomers use a unit called a light-year for measuring very large distances. One light-year is equal to the distance that light travels in one year (about 6 trillion miles). Therefore, Proxima Centauri is about 4.2 light-years away from earth. Rigel is a star that is about 810 light-years away. And stars in the Andromeda galaxy are over 2 million light-years away. Some stars are billions of light-years away.

Color, size, and relative motion

Color is the third characteristic of stars. Some stars are blue, or bluish-white, while others are yellow, orange, or even red. The color of the star is determined by the surface temperature of the star. Blue stars are the hottest, with a surface temperature believed to be approximately 54,000°F (30,000°C). White stars have a surface temperature of about 20,000°F (11,000°C), while the cooler red stars have a surface temperature that is only about 5,400°F (3,000°C). Our sun is a yellow star and is about 11,000°F (6,000°C) on the surface. Orange stars are about 7,600°F (4,200°C).

> ## FUN FACT
> Sirius emits about 25 times as much light as our sun. Rigel emits about 50,000 times as much light as our sun. But Sirius appears brighter to us than Rigel because Sirius is only 8.8 light-years away and Rigel is 810 light-years away.

*earth – moon
239,000 miles*

*earth from sun
93 million miles*

Stars vary in size as much as they vary in color. In general, the hotter and brighter stars are bigger and more dense. This is not a hard and fast rule, however, and some smaller stars are hotter than some larger stars.

Finally, stars are described by their motion through space. This measurement does not refer to the movement of the stars through the sky each night or each season, but the movement of stars relative to other stars. Because of the immense distances to the stars, they do not appear to move much, even over hundreds of years. However, some stars that were recorded 1,500 years ago by a Greek named Ptolemy, who made very accurate measurements, have changed location by as much as one degree, or two times the width of the moon.

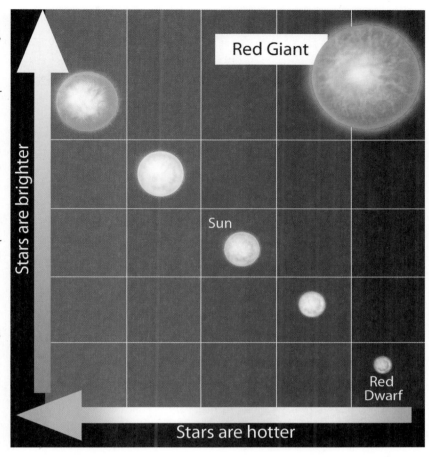

Whether the stars are near or far, bright or dim, we can enjoy them and recognize the signature of the Great Designer as we look to the skies and see the beauty there. ■

 # SIMULATING STARLIGHT

Complete the "Starlight" worksheet.

WHAT DID WE LEARN?

- What is the unit of distance used to measure how far away something is in space?

- How far is a light-year?

- What does the color of a star tell us about that star?

TAKING IT FURTHER

- What causes stars to appear to move in the sky?

- How can we determine if a star's absolute distance from the earth is actually changing over time?

- Why is brightness not a good indicator of the distance of a star from the earth?

ORIGIN OF STARS

Where do stars come from? You may get different answers to this question depending on who you ask. Many evolutionists believe that stars are formed in gas clouds such as nebulae, where the particles of gas and dust clump together and eventually become so dense that they form stars. Pictures of bright areas within nebulae have been used to support this hypothesis. However, the truth is that no scientist has ever witnessed the birth of a star. Particles do not just clump together and become more dense without some outside force pushing them together. What is actually observed inside nebulae is gas that is expanding not contracting. Many scientists will admit that they do not really understand the process by which stars are formed.

So where did the stars come from? According to the Bible, God created the stars on Day 4 of the Creation Week. Psalm 8:3–4 says, "When I consider Your heavens, the work of Your fingers, the moon and the stars, which You have ordained, what is man that You are mindful of him, and the son of man that You visit him?"

Which makes sense to you? To believe the Bible or to believe men who don't really understand the process of star formation and have never witnessed a single star being formed? Take a few minutes and make a couple of calculations that may help you appreciate the situation. Scientists estimate that there are about 100 billion galaxies and that each galaxy contains an average of 200 billion stars per galaxy. How many

stars do scientists estimate are in the universe? If you did your calculations right you should have come up with a number that is 2 followed by 22 zeros. Evolutionists believe that the universe is about 14 billion years old. How many stars would have to be formed each year in order for all the stars we have today to exist? This would be 1.4 trillion stars per year. Divide that number by 365 days per year and what do you get? That would be about 4 billion stars formed per day, every day for the past 14 billion years. Yet today, scientists do not observe even one new star forming.

This brings up another question. How long does a star live? A star is continually using up energy and slowly burning out. The more massive a star is, the shorter its lifespan. This may seem backwards to you. But the more mass a star has, the more gravity pulls on that mass and the hotter it burns, so it will use its fuel up faster than a less massive star, which burns slower. The brightest stars we see today will burn out in about 10 million years. If the universe is 20 billion years old, how can we still see so many bright stars today? Evolutionists claim that new stars are being born to replace the dying stars, yet we do not actually see this taking place. We can believe the Bible when it says that God formed the stars at the creation of the world only a few thousand years ago.

Many evolutionists believe that stars form in nebulae, like the Orion Nebula pictured here.

HEAVENLY BODIES

More than just stars

LESSON 7

What kinds of objects do we find in the universe?

Words to know:

nova

supernova

· nebula

BEGINNERS

We live on the planet earth, which orbits a star that we call the sun. All of the planets and moons that orbit the sun are part of our solar system. But our sun is only one star. Stars are often found in large groups, which are called galaxies. Our sun is part of a large group of stars called the Milky Way galaxy. Stars in a galaxy all move around a central point. Some galaxies, or groups of stars, form an oval shape; others form a spiral or pinwheel shape.

Scientists believe there are billions of galaxies in the universe with billions of stars in each galaxy. That is a lot of stars! But there are other interesting things in space besides stars. There are clouds of gas and dust that can be seen in space. These clouds are called nebulae. Bright nebulae are found near stars, and dark nebulae are far away from stars. We are going to learn about planets, asteroids, comets, and meteors in other lessons. The universe is a truly beautiful and interesting place, and it shows God's majesty.

- What is a galaxy?
- What is the name of our galaxy?

Objects in the universe can be very far apart. Astronomers have measured star systems that are 100,000 or even 1 million light-years away from earth, and they believe that other systems are billions of light-years away. Some people argue that the earth must be millions or even billions of years old in order for light from so far away to reach here. Some have speculated that the earth looked mature when it was first created, and so the starlight was created in place to reach the earth. Recently, however, a new theory has been suggested that says that due to the intense gravity at the center of the universe when it was created, time moved more slowly in earth's reference frame (predicted by General Relativity), allowing "billions of years" to pass in the outer reaches of the expanding universe while only six days passed on earth. Research into this "problem" continues, but in any case, light from the far reaches of the universe does not prove the earth to be millions or billions of years old. (See www.answersingenesis.org/go/starlight for more information.)

Elliptical galaxy

Spiral galaxy

Stars

The most obvious objects in the universe are the stars. Groups of stars that appear to move together are called star clusters. Some clusters of stars are grouped closely and are called globular clusters. Other stars move together but are spread out and are called open clusters. Star clusters usually consist of thousands of stars.

Larger groups of stars are called galaxies. A typical galaxy consists of millions or billions of stars rotating around a center. The most common shape for a galaxy is elliptical. Other galaxies are spiral shaped like the one shown bottom left, and others have irregular shapes. Our sun is part of the Milky Way galaxy. The Milky Way is believed to be about 100,000 light-years across.

Astronomers have discovered many unusual stars within these clusters and galaxies. Some stars appear to increase and then later decrease in brightness. One kind of star that varies in brightness is a Cepheid variable star. This is a star that expands and becomes brighter, and then contracts and becomes dimmer on a regular schedule of days or weeks. Varying brightness can also be caused by eclipsing binary stars, where two stars revolve around each other. When both stars are side by side they appear as one very bright star, but when one star is in front of the other, they appear to be a single dimmer star.

Another unusual type of star is called a nova. A star becomes a nova when it experiences an explosion. The brightness of the star may increase by as much as ten times. This increased brightness may last for months. Then the star goes back to its normal brightness. It is believed that the star loses some of its mass due to the explosion. If a star has an unusually large explosion, it is called a supernova. A supernova may become as much as 20 times brighter, and the explosion may destroy the star. In 1604, records show that a star became so bright that it was visible during the day—this is believed to have been a supernova.

When a massive star explodes, its core collapses into a neutron star. This is an extremely dense, small object. Neutron stars spin very rapidly, and emit pulses of radio waves. (If earth is lined up with these pulses, the neutron star is called a *pulsar*.) Some stars are so massive when they explode their core is crushed into a tiny point. The area around this point is called a black hole. Scientists cannot see a black hole, but they see the effects of it. They see a very dense area pulling gas into itself and emitting x-rays. The gravity in a black hole is so strong that nothing can escape it, not even light.

Nebulae

Stars are not the only interesting objects in space. There are also planets, asteroids, comets, and meteors. We will examine each of these in more detail in following lessons. Another interesting object is a nebula. A nebula is a cloud of gas and dust that is in space. If there are stars near the nebula, it is called a bright nebula because it glows hot from the light from nearby stars. The picture at the beginning of this lesson is a bright nebula. If there are no stars nearby it is called a dark nebula. Dark nebulae can only be seen because they block out a part of the sky like a shadow. Some nebulae are believed to be the remains of supernovas. The Crab Nebula was formed when a star exploded in AD 1054. Most nebulae have an irregular shape, but sometimes they can form something that resembles a known object. One dark nebula is called the Horsehead Nebula because it resembles a horse's head. One type of nebula, a planetary nebula, has a ring or disk shape. Planetary nebulae are rings of gases expanding outward from a central hot star.

Quasars

One final unusual object in space is a quasar. Quasars are relatively small objects that are as bright as an entire galaxy. They appear faint because they are very far away. Some think they might be the centers of some distant galaxies with a supermassive black hole at the center causing some sort of energetic action. They appear

MAKING A NEBULA

Astronomers can see bright nebulae with telescopes because they glow hot from the light of neighboring stars. However, detecting dark nebulae is a little more difficult. Astronomers know where a nebula is, even if there are no nearby stars, because it blocks out the light from the stars that are behind

it. To demonstrate this, perform the following activity.

Purpose: To see how dark nebulae are identified

Materials: flashlight, pencil

Procedure:

1. Shine a flashlight on the wall.

2. Have someone hold a pencil or some other object in the

beam of light and observe the shadow on the wall.

Conclusion:

The shadow indicates that something is blocking the light, even if you can't distinctly see it. The shadow of the pencil shows us its shape, just as the area blocked by a nebula shows us its shape.

to be moving away from the earth at very high speeds. There is much controversy regarding these mysterious objects.

God designed the universe to be an astounding place. It is bigger than we can imagine. It has more stars than we can possibly count. And it supplies us with unlimited hours of fascination and research possibilities. So, enjoy studying our universe. ■

WHAT DID WE LEARN?

- What is a cluster of stars?
- What is a galaxy?
- Explain the difference between a nova, a supernova, and a neutron star.

TAKING IT FURTHER

- How can a star appear to become brighter and dimmer on a regular basis?
- Why does starlight from millions of light-years away not prove that the earth is old?

DISTANT STARLIGHT

The most popular evolutionist idea for the origin of the universe is called the big bang theory. According to this theory, about 15–20 billion years ago the universe came into existence when space began to rapidly expand from a tiny point, called a singularity. Within this expanding mass, planets, moons, stars, and other celestial bodies were formed.

There are several problems with this idea that scientists have not been able to adequately address. First, where did all of the matter and energy in the universe come from to begin with? Second, what caused the singularity to begin to expand? Third, if stars have been forming and dying for billions of years, why don't we see stars forming today? Fourth, how did planets, stars, and other celestial bodies form from nothing but hydrogen gas—the first element in the big bang? The laws of science tell us that particles would

be spread out in all directions, and not come together to form planets and stars.

The Bible clearly states that God created the universe and everything in it. However, there is one important question that creationists are trying to answer. If the universe was created only a few thousand years ago, how can we see light from a star that is millions of light-years away? It seems that the light from that star would have originated millions of years ago, yet the Bible says that everything was created only a few thousand years ago. Although creationists disagree on what is the best answer to this "problem," there are several.

One theory proposes that the speed of light was faster in the past than it is today, so light would cover more distance in less time. This would mean that a star would not have to be millions of years old for its light to reach the earth. Not everyone agrees that

this is a good explanation, but some people say it is possible. Another theory uses Einstein's theory of relativity that explains that time is not constant but is affected by gravitational pull. Thus, if the earth is near the center of the universe, a few thousand years may have passed here, while millions of years passed at the edge of the expanding universe. During the Creation Week, while six days passed from the perspective of earth, billions of years could have passed in the far reaches of space. This theory is supported by many of the observations that we see, but is still being investigated.

If scientists cannot explain how we see distant starlight, does this mean the Bible is wrong? No. There have been, and there still are, many areas of science that cannot be adequately explained. Yet, the Bible is God's Word and can be trusted.

ASTRONOMY VS. ASTROLOGY

Astronomy is the study of the objects in the universe. It is the study of planets, moons, stars, comets, and other heavenly bodies. Astronomers seek to observe and understand what these objects are made of, how far away they are from the earth, how they function, and how they affect each other. Astrology, on the other hand, is not science, but is based on superstition and on occult practices. Understanding the movement of the stars and using the constellations as a navigation tool is science—Astronomy. But using the stars to predict the future is superstition—Astrology.

Astrology has its roots in ancient religions including the beliefs of the Babylonians, Egyptians, and Indians. Ancient people believed that heavenly bodies such as the moon, planets, and stars gave off vibrations that could affect the future of particular people. They used the alignment of the stars and planets to predict what would happen in the future. When scientists discovered how far away these heavenly bodies are, it became obvious that their vibrations could not have any effect on people on earth. However, these superstitious ideas still persist. Many people today still believe in horoscopes, which are predictions based on the movement of the stars.

Astrologers make predictions about the futures of other people. Many times these predictions are made after seeking information from a spirit guide. This is witchcraft and should be avoided by all Christians. The Bible says, "There shall not be found among you anyone ... who practices witchcraft, or a soothsayer, or one who interprets omens, or a sorcerer, or one who conjures spells, or a medium, or a spiritist, or one who calls up the dead. For all who do these things are an abomination to the LORD" (Deut. 18:10–12). Using the stars to try and predict the future is obviously against God's will.

God intended the heavens to be a source of joy and wonder. Psalm 19:1 says, "The heavens declare the glory of God; and the firmament shows His handiwork." And Job 38:33 says, "Do you know the ordinances of the heavens? Can you set their dominion over the earth?" We should study to understand the way the universe works, but we should not be superstitious in our study of the heavens. We should instead look for ways to glorify God in our study of the universe He created.

ASTEROIDS

Minor planets

LESSON

8

What is an asteroid and where are most of them located?

Words to know:

asteroid

asteroid belt

BEGINNERS

What do you think an asteroid might be? If you think it is something found in space, you are right. An **asteroid** is a large rock that orbits the sun. An asteroid is too small to be a planet, but it is big enough to go around the sun. Most asteroids are between 1 and 600 miles across.

Most of the asteroids in our solar system can be found orbiting the sun in between Mars and Jupiter. This area is called the **asteroid belt**. There are many thousands of asteroids, but only about 3,000 of them are big enough to receive a specific name.

- What is an asteroid?

- Where is the asteroid belt located?

Relative to the distance between other orbits, there is a large gap between the orbit of Mars and the orbit of Jupiter. However, this gap is not empty. It is filled with what many call the minor planets. These are large pieces of rock that orbit the sun, just as the planets do. Because when they were first discovered they looked star-like to the observers, they were called asteroids, meaning star-like. This area between Mars and Jupiter is called the asteroid belt.

We know today that the asteroid belt contains millions of chunks of rock in a regular orbit around the sun. Most of the asteroids are small, but a few are fairly large. The first and largest asteroid was discovered in 1801, and is called Ceres. Ceres is about 600 miles (965 km) in diameter. Ceres is now considered a dwarf planet. Other large asteroids were discovered between 1804 and 1807. Pella and Vesta are two asteroids that are over 300 miles (483 km) across and were discovered during this time. These may also be classified as dwarf planets in the future. By 1890, with the improvement of telescopes, over 350 minor planets had been identified. Today, about 3,000 minor planets have been named. The rest of the asteroids are considered too small to warrant the name "minor planet."

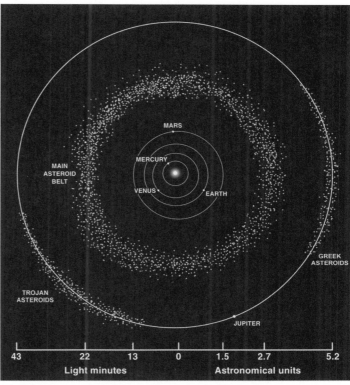

Asteroid families

Asteroids are objects that are smaller than planets yet circle the sun in a regular orbit. Most of the asteroids in our solar system are found in the asteroid belt. However,

NAMING ASTEROIDS

The first person to discover an asteroid in the asteroid belt was an Italian astronomer named Giuseppe Piazzi. He named his discovery Ceres after the Roman goddess of grain and agriculture. Later discoveries were also given the names of Greek or Roman goddesses. But there were many more asteroids than there were goddesses to name them after, so scientists began to name them after famous people. The asteroid Piazzi was named after the astronomer who discovered the first asteroid. Some asteroids were named after cities, flowers, and even pets. Pretend that you are an astronomer that has just discovered five new asteroids. What would you name them? Write down your names on a piece of paper. Remember that most asteroid names have been changed to make them sound feminine. For example, Pittsburghia was named after Pittsburgh with an "ia" added to the end.

some groups of asteroids (called families) have different orbits. A group of asteroids called the Amor Asteroids has an orbit that crosses the orbit of Mars. Two other families, the Trojan and Greek Asteroids, are in the same orbit as Jupiter.

The Apollo Asteroids cross the orbit of the earth, but are located in a different plane, so they will not come in contact with the earth. A few asteroids have orbits that come relatively close to the earth. These asteroids are nicknamed "earth grazers." The closest recorded asteroid came within 84,000 miles (135,000 km) of the earth in 1993.

All asteroids are relatively small. Because of their small mass, they do not have enough gravity to hold an atmosphere, so they are all lifeless rocks. Still, they are interesting objects to study. At one time it was suggested that the rocks in the asteroid belt may have been a planet that broke apart. But there is no evidence for this and no method that explains how a planet could have broken up and leave its pieces in its original orbit. Besides, if all the asteroids in the asteroid belt were put together, they would still be smaller than the earth's moon—not a very impressive planet. ■

WHAT DID WE LEARN?

- What is an asteroid?
- Where are most asteroids in our solar system located?
- What is another name for asteroids?

TAKING IT FURTHER

- What is the chance that an asteroid will hit the earth?

TROJAN ASTEROIDS

When a group of asteroids travels in the same path, especially if it is outside the asteroid belt, the group is called a family. As mentioned earlier, the Trojan family of asteroids travels in the same orbit as the planet Jupiter. All of the asteroids in this family are named after the heroes of the Trojan War. In general, the asteroids shown to the right of Jupiter in the diagram are named after Greek heroes and those to the left of Jupiter are named after the heroes from Troy. Do some research and find out who these heroes were and what they did during the Trojan War. This war is chronicled in the book *The Iliad* by Homer.

COMETS

Look at that tail!

LESSON

9

What is a comet made of and how does it travel?

Words to know:

comet

BEGINNERS

Most of the bright spots you see in the sky are stars, but some of them may be planets or other objects in space. You have learned that asteroids are large rocks that orbit the sun. Comets are another kind of object that orbits the sun. **Comets** are different from asteroids because they are made mostly of ice with dust and bits of rock mixed in.

As a comet gets closer to the sun, it begins to vaporize. The ice turns into gas, and some of the dust particles are pushed away from the comet. This makes the comet look like a bright ball with a tail. The bright part of the comet is called the head, and the dust particles that are flowing away from the sun form the tail.

- What is a comet?

- What are the two parts of a comet?

W hat has a tail when it is near the sun, but loses it when it is far away? A comet does. Comets are another interesting phenomenon seen in space. Just like asteroids, comets have a regular orbit around the sun. But unlike asteroids, comets are made mostly of bits of rock and dust surrounded by ice. Comets orbit the sun once every ten to several hundred years. The time it takes to orbit the sun is called the comet's period. Because comets are seen so seldom, they were not always recognized as being the same object from one sighting to the next.

Edmond Halley

Edmond Halley (1656–1742) was the first scientist to apply the laws of astronomy to comets. Using Newton's law of gravity, he predicted the orbits for several comets. He found that comets that had appeared in 1531, 1607, and 1682 all had nearly identical orbits. So he concluded that they were actually the same comet. He predicted that the comet would appear again in 1758. He died before that time. However, when the comet did show up as he predicted, the scientists named the comet Halley's Comet. It appears every 75–76 years. It was last seen in 1986. Halley's Comet has a very elliptical, or stretched, orbit that reaches nearly to Pluto.

Comets have two main parts: the head and the tail. The head of the comet has a nucleus and a coma. The nucleus contains most of the mass of the comet. It is made up of bits of rock and dust that are frozen in a ball of ice. The nucleus of most comets is less than 50 miles (80 km) in diameter. Halley's Comet has a nucleus that is less than 10 miles (16 km) across.

Surrounding the nucleus is the coma. The coma is a cloud of material that has been ejected from the nucleus. It contains gases and ice particles. It can be very large. It reflects light very well and is what allows astronomers with telescopes, and sometimes even observers without telescopes, to be able to see the comet when it passes close to the earth.

The tail of the comet consists of gases and dust particles that are forced back from the nucleus by solar winds and the pressure of sunlight. Comets may have one or two types of tails. A Type I tail forms quickly and is made mostly of gases that are forced back by the solar winds. A Type I tail goes straight out from the nucleus. A Type II tail is comprised mostly of dust that is pushed by

FUN FACT

In 1999 astronomers discovered 91 new comets. Six of these were discovered by amateur astronomers. So keep looking to the sky—you may discover something new.

FUN FACT

Project Deep Impact launched a space vehicle on December 30, 2004 that impacted the Comet Tempel 1 on July 4, 2005. This impact created a large crater that allowed scientists to study the surface and interior of the comet.

COMET MODEL

Purpose: To make a model of a comet

Materials: Styrofoam ball, tag board, glue, glitter

Procedure:

1. Cut a small Styrofoam ball in half.

2. Glue it to a piece of tag board as the nucleus of the comet.

3. Using glue and glitter, add a coma around the nucleus and one or more tails to the comet. Be sure to remember that the tail points away from the sun.

the radiation pressure from the sun. It forms slowly and curves away from the main tail. The tails of a comet always point away from the sun regardless of whether the comet is moving toward or away from the sun. As the comet gets farther from the sun, its tail shrinks, and when it is far from the sun it does not have a tail at all.

Some comets have only one type of tail. Some have both types. Some comets even have several of each type of tail. The Great Comet of 1744 had six tails! A second tail is slightly visible in the photograph here.

Comets are fragile and can break apart when they are approaching the sun. Break-up can be rapid. Most comets last for only a few hundred to a few thousand years. Evolu-

Two-tailed comet

tionists cannot explain the existence of some short-period comets. If the universe is actually billions of years old, all of these comets should have completely vaporized by now. Some scientists try to explain comets by saying there is a storehouse or birthplace for comets beyond Pluto, called the Oort cloud, but there is no evidence to support this idea. Discovery of previously unknown comets is due to better telescopes and more astronomers searching the skies. God designed comets and put them in place at creation, and we can enjoy them today. ■

WHAT DID WE LEARN?

- What is a comet?
- Who was the first person to accurately predict the orbit of comets?
- What are the two main parts of a comet?

TAKING IT FURTHER

- Why does a comet's tail always point away from the sun?
- Why doesn't a comet have a tail when it is far from the sun?
- When will Halley's Comet next appear?

GOD CREATED COMETS

Comets can be powerful tools to support the Bible. Explain how what you have learned about comets can support each of the following Scriptures:

1. "Lift up your eyes to the heavens, And look on the earth beneath. For the heavens will vanish away like smoke, The earth will grow old like a garment, And those who dwell in it will die in like manner; But My salvation will be forever, And My righteousness will not be abolished." Isaiah 51:6

2. Then God said, "Let there be lights in the firmament of the heavens to divide the day from the night; and let them be for signs and seasons, and for days and years." Genesis 1:15

3. Thus says the LORD: "Do not learn the way of the Gentiles; Do not be dismayed at the signs of heaven, For the Gentiles are dismayed at them." Jeremiah 10:2

4. "In the beginning was the Word, and the Word was with God, and the Word was God. He was in the beginning with God. All things were made through Him, and without Him nothing was made that was made." John 1:1–3

FUN FACT

The NASA *Stardust* spacecraft returned from a successful mission on January 15, 2006 after retrieving actual particles of gases and dust near the head of comet Wild 2. These are being studied by scientists to help us understand the composition of comets.

Meteors

Shooting stars

LESSON 10

What is the difference between a meteor, a meteorite, and a meteoroid?

Words to know:

meteor

meteoroid

meteorite

Beginners

Have you ever seen a shooting star, something that looked like a bright light streaking across the sky and then disappearing? What you saw was probably a meteor. A **meteor** is a rock or other object in space that has gotten close enough to the earth to be pulled down by gravity. As the rock goes through the earth's atmosphere it gets hot and burns up. That is why it is so bright.

Many meteors are the remains of comets that have broken apart. In the last lesson, you learned that comets disintegrate a little every time they pass the sun. Eventually, comets get small enough that they break apart, leaving bits of dust, rock, and ice floating in space. When the earth passes near the area where the comet broke up, some of the pieces are pulled into the earth's atmosphere and you might see a meteor shower, which is many shooting stars in a short period of time.

- **What is a meteor?**

- **Where do many meteors come from?**

Have you ever watched the night sky and saw what appeared to be a star streaking across the sky? You may have said you saw a shooting star or a falling star. What you probably saw was a meteor. Meteors are objects from space that have been pulled into the earth's atmosphere by gravity. Friction from the atmosphere heats the object up, usually burning it up before it reaches the surface of the earth. Most meteors are visible at 60–80 miles (95–130 km) above the surface of the earth. Some meteors have been clocked at speeds up to 45 miles per second (72,000 m/s)!

Particles in space that are too small to be an asteroid or a comet are called meteoroids. Occasionally, an object comes close enough to the earth to get caught in the earth's gravitational pull. If the object enters the earth's atmosphere, but does not reach the ground, it is called a meteor. If the object is large enough that at least some of it survives the hot trip through the atmosphere and hits the surface of the earth it is called a meteorite. Only about 1 out of 1,000,000 meteors becomes a meteorite. Our atmosphere is part of God's provision to protect us from the debris in space.

Your chances of seeing a meteor depend on many things. You have to be in the right place at the right time and look at the right part of the sky. However, you can increase your chances of seeing a meteor by watching the sky on particular days and at certain times of the night. Beginning in late July and peaking around August 12th of each year, the earth passes through the remains of the tail of a comet called Comet Swift-Tuttle. These meteors appear to move away from the constellation Perseus, and are thus called the Perseids. A similar meteor shower is experienced around mid-November each year, as the earth passes through the remains of the tail of Comet Tempel-Tuttle, which passes through the earth's orbit every 33 years. These meteors are called the Leonids because they always appear to streak away from the constellation Leo.

FUN FACT

There are many craters in various locations around the earth that are believed to be the result of the impact of meteorites. The most famous crater is the Barringer Meteorite Crater near Winslow, Arizona, shown in this picture. It is 4,150 feet (1265 m) across and 570 feet (174 m) deep. The rim of the crater rises 150 feet (46 m) above the surrounding terrain. Scientists believe that the meteorite that struck the area was about 100 feet (30 m) in diameter. They believe that it exploded on impact, causing the giant crater. About 30 tons of fragments have been unearthed, but no large pieces of rock have been found.

Every September for the past several years scientists from NASA have conducted experiments near this crater because the harsh conditions there are similar in many ways to the conditions on the moon and Mars.

THE SKY IS FALLING

Purpose: To demonstrate how impact craters form

Materials: pie pan, flour, toys, salt, marble, golf ball

Procedure:

1. Fill a pie pan half full of flour.

2. Arrange toys to represent a town.

3. Sprinkle salt over the town and observe any damage.

4. Drop a marble on the town from a height of two feet and observe any damage.

5. Finally, drop a golf ball on the town from a height of two feet.

Questions:

• Did the salt do any damage? This represents the vast majority of the meteorites that hit the earth. Most are so small they fall as dust and are not even noticed.

• How much damage did the marble cause? This represents a few larger meteorites that occasionally strike the earth, causing some damage.

• How much damage did the golf ball do to the town compared to the marble? Did it make a crater in the flour? This represents the very rare large meteorite that can do substantial damage to the earth.

You also have a better chance of seeing meteors if you look at the sky after midnight. After midnight, your part of the earth is on the leading edge as the earth moves around the sun. Thus, your part of the earth is approaching meteoroids rather than moving away from them. This phenomenon of observing more meteors after midnight is one of the proofs that Copernicus and others used to prove that the earth is moving through space rather than being stationary.

Most meteorites are very small by the time they reach the surface of the earth. But occasionally a large meteorite manages to strike the earth's surface. The largest meteorite ever discovered is the Hoba meteorite, found in Namibia, Africa in 1920. It weighs almost 70 tons and is so large that no one has attempted to move it from where it landed. Studies have shown that most meteorites are made from silicates and other stone. Most of the remaining meteorites are made from iron. Rarely, meteorites are found to contain stone and iron in equal proportions.

Even though meteorites are relatively rare, if the earth were billions of years old, we would expect to discover hundreds of meteorites in the fossil layers of the earth. However, there have only been a few confirmed discoveries of meteorites in the fossil layers. If, as the Bible says, the earth is only a few thousand years old and most fossils are a result of the great Flood, we would not expect to find many meteorites in the fossil layers. ■

WHAT DID WE LEARN?

• What is the difference between a meteoroid, meteor, and meteorite?

• When is the best time to watch for meteors?

TAKING IT FURTHER

• Space dust (extremely small meteorites) is constantly falling on the earth. If this has been going on for billions of years, what would you expect to find on the earth and in the oceans?

• Have we discovered these things?

WHAT HAPPENED TO THE DINOSAURS?

Outer Space

Did a meteor really cause the extinction of the dinosaurs? Many science books today suggest that a very large meteor or comet could have collided with the earth millions of years ago. They suggest that this collision put tons of debris into the air causing the climate to change significantly and the dinosaurs were unable to adapt to these changes and thus died out. But is there any evidence for this kind of an event? A collision of this magnitude would have left a gigantic crater; however, even though a few large craters have been found around the world, none has been found that is nearly large enough to change the climate on a global scale. And if the dinosaurs went extinct because of this global catastrophe, why did many other kinds of animals survive this event with no problems.

Other astronomical events have also been suggested for the demise of the dinosaurs. Some people suggest that a passing comet could have poisoned the air, or a meteorite may have landed in the ocean, causing a giant wave that washed all life out to sea. Another suggestion is that a nearby supernova exploded, poisoning the earth with deadly radiation. None of these ideas has any significant evidence to support it.

What does the Bible say about dinosaurs? You probably won't find the word dinosaur in a concordance, but you can read the following verses and get an idea of what God's Word says about the fate of dinosaurs. Read Genesis 6:19–7:5, Genesis 8:18–19, Job 40:15–41:10. These verses indicate that representatives of the dinosaurs would have been on the Ark with Noah and were saved from the Flood, but that the rest were killed by the floodwaters. The verses in Job indicate that at least some of the dinosaurs were still alive during Job's life, so some of the dinosaurs continued to live after the Flood. However, it seems that most dinosaurs did not cope well with the changed earth after the Flood and became extinct like many other animals have in the past. For more on dinosaurs and how to explain them using the Bible, see www.answersingenesis. org/go/dinosaur.

UNIT **3**

SUN & MOON

OVERVIEW OF OUR SOLAR SYSTEM

Revolving around the sun

Jupiter
Uranus
Nept
Saturn

LESSON

11

How many planets do we have in our solar system?

Challenge words:

elliptical

perihelion

aphelion

BEGINNERS

Our solar system includes the sun and everything that moves around it. You have already learned about asteroids and comets that orbit our sun, but now you are going to learn about the bigger things in the solar system. This includes eight planets, several moons, and the biggest thing in our solar system, the sun.

There are four planets that are relatively small and four planets that are much larger. The sun is the center of our solar system. The four smaller planets that are closest to the sun are Mercury, Venus, Earth, and Mars. Then comes the asteroid belt, which includes the dwarf planet Ceres. Moving farther away from the sun, we come to the four large planets, Jupiter, Saturn, Uranus, and Neptune. Farthest away from the sun are the dwarf planets Pluto and Eris. We will learn more about each of these objects, as well as the sun and the moon, in the next lessons.

- How many planets are in our solar system?

- What are the names of the small planets?

- What are the names of the large planets?

The universe is so vast and most objects in it are so far away that it is difficult for us to study them in much detail. Because of this, we are most familiar with the objects in our own solar system. Our solar system is the collection of objects that revolve around our sun. Our sun is one of millions of stars in the Milky Way Galaxy. It is located about ⅗ of the way from the center of the galaxy, in between two of the pinwheel arms.

Our solar system consists of one star (the sun), eight planets, several dwarf planets, many moons, asteroids, comets, and meteoroids. The sun is the largest and most massive object in our solar system. It is a yellow star. Its gravitational pull is what keeps all the other objects in their orbits around it.

Mercury is the closest planet to the sun. It is a dead planet. It is sun-baked and extremely hot. The surface resembles the surface of the earth's moon. The second planet from the sun is Venus. Venus is covered with a thick layer of yellow clouds made of sulfuric acid. The third planet out from the sun is earth. Earth is the only planet in our solar system that is able to support life. God designed this planet just for us! The fourth planet is Mars. Mars has a reddish color due to the iron oxide (rust) in its soil. It also has two tiny moons that orbit it.

These first four planets are called the terrestrial, or earth-like, planets because they are all composed of rocky material. The first four planets are also called the inner planets. Their orbits are all relatively close to the sun. Between Mars and the rest of the planets is a large gap. As you learned in lesson 8, this gap is called the asteroid belt and is filled with asteroids. There are millions of small chunks of rock in this belt and thousands of larger rocks. The largest asteroid in the asteroid belt, Ceres, is about one-third the size of the earth's moon and is now classified as a dwarf planet.

Outside the asteroid belt are the four outer planets: Jupiter, Saturn, Uranus, and Neptune. These are also called the Jovian, or Jupiter-like, planets. This means that, like Jupiter, these four planets are all made of gas. They are not solid like the inner planets. Pluto was considered a planet until 2006 when it was demoted to the status of a dwarf planet.

Of the outer planets, Jupiter is closest to the sun and is the largest planet in the solar system. Saturn is the sixth planet from the sun and the second largest planet.

Sun & Moon

Relative sizes of the sun and planets

FUN FACT

With improvements in technology and increased interest in astronomy, new discoveries about our solar system are constantly being made. For example, in January 2003, scientists announced the discovery of three new moons around the planet Neptune. This was the first discovery of new moons made by a land-based telescope since 1949. Many other new discoveries have been made since the launching of the Hubble Space Telescope. In July 1997, astronomers discovered the most distant galaxy to date. It is 13 billion light-years from earth. In August 1997, astronomers discovered what is believed to be the first observed planet outside our solar system. And in July of 2005, astronomers announced the discovery of what was originally thought to be a tenth planet in our solar system, but is now classified as a dwarf planet called Eris. Eris is 30% more massive than Pluto and orbits nearly three times as far from the sun.

Saturn is most well known and recognized for the thousands of beautiful rings that surround it. The seventh planet is Uranus. Uranus also has a few rings. The eighth planet from the sun is Neptune, a bluish/green gas planet.

Beyond the orbit of Neptune are Pluto and Eris, classified as dwarf planets. Because these worlds are so far away and so small, and because no space probes have reached them yet, we do not know very much about them. They are believed to be terrestrial and composed of a combination of frozen methane, water, and rock.

In addition to the sun and planets, our solar system contains many moons. Several of the planets have moons revolving around them. Some of the planets have multiple moons. There are also many comets and groups of asteroids that orbit the sun. Some of these have very elongated orbits and are only rarely visible from the earth.

The outer planets are much larger than the inner planets. One beneficial effect of this for earth is that the larger planets have larger gravitational pulls and thus capture many of the meteoroids that might otherwise make it to earth's orbit. This is another example of God's wonderful design for the universe. ■

LEARNING THE NAMES OF THE PLANETS

You can learn the names and orders of the planets (including Pluto, the former planet) by learning the following song—sung to the tune of "Twinkle, Twinkle Little Star."

Mercury, Venus, Earth and Mars
These are the planets that dwell with the stars.
Jupiter, Saturn, Uranus, too

Neptune, Pluto, I know them do you?
Mercury Venus, Earth and Mars
These are the planets that dwell with the stars.

WHAT DID WE LEARN?

- Name the eight planets in our solar system.
- Name two dwarf planets.
- Which planets can support life?

TAKING IT FURTHER

- What are the major differences between the inner and outer planets?
- Why are the gas planets called Jovian planets?

LAWS OF PLANETARY MOTION

In lesson 2, we mentioned that Johannes Kepler discovered that the planets move in an **elliptical**, or squashed circle, path around the sun. This was an important discovery because it explained the observations made by many astronomers and showed that God's design was a good and orderly design. Before Kepler's discovery, many people, including Copernicus and Tycho Brahe, struggled to explain the apparent course for Mars. It appeared that Mars moved backward at times with respect to the earth. This was explained by the fact that earth is moving more quickly through its orbit than Mars is and therefore catches up with Mars and passes it, making Mars appear to move backward (see below). Although Copernicus believed this was the case, he and others could not find a correct path for Mars.

Tycho Brahe made thousands of very accurate measurements of the location of Mars but could not plot its exact course. Then, near the end of Brahe's life, God arranged for Kepler to join him in his work. Kepler had been living in Austria, but when a persecution of Protestants broke out, Kepler left Austria and moved to Prague. There he went to work for Brahe. After many years of work, Kepler finally discovered that the planets move in ellipses.

Through all his work, Kepler discovered three laws that affect the movements of not only the planets, but also the moons and man-made satellites as well. Below are Kepler's three laws of planetary motion:

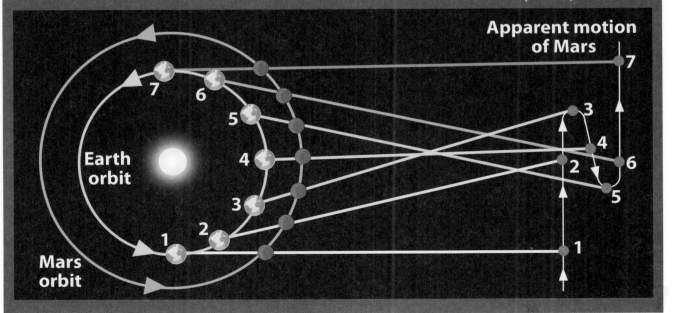

Apparent motion of Mars

Earth orbit

Mars orbit

Sun & Moon

1. Planets move in ellipses with the sun as the focus.

2. An imaginary line from the center of the sun to the center of a planet always sweeps over an equal area in equal time.

3. The squares of the periods of the planets are proportional to the cubes of their distances from the sun.

These may sound confusing so let's look at what each of these laws means. The first law is pretty straightforward; planets travel in ellipses around the sun. The second law means that when the planet is closer to the sun it moves along its orbit faster, or covers more distance, than it does when it is further away from the sun. The third law means that the time it takes for a planet to make one orbit around the sun increases the farther away the planet is from the sun.

Because the planets do not travel in a circle around the sun, they are not always the same distance from the sun. When a planet is at its closest to the sun it is said to be at its **perihelion** (per uh HEE lee un), and when it is at its farthest point from the sun it is said to be at its **aphelion** (au FEE lee un). Charts that show distance from the sun usually show the average distance the planet is from the sun. Venus and Neptune have orbits that are nearly circular, so their distance from the sun does not vary as much as planets with more stretched orbits. Pluto has a very eccentric or stretched orbit, and its actual distance from the sun may be very different from its average. Also, Pluto's orbit crosses Neptune's orbit, so sometimes Pluto is the farther from the sun and other times Neptune is the farthest from the sun. The orbits of all eight planets are in nearly the same plane around the sun. The dwarf planet Pluto orbits at about a 17° tilt compared to the rest of the planets, and the orbit of Eris is tilted by 44°.

Purpose: To demonstrate how Neptune's and Pluto's paths can cross.

Materials: piece of paper, tape, cardboard, two thumb tacks, string, pencil

Procedure:

1. Tape a piece of paper to a piece of cardboard.

2. In the center of the paper place two thumb tacks 3 inches apart horizontally.

3. Cut a piece of string 12 inches long and tie it in a circle and place the string around the tacks.

4. Place your pencil against the inside of the string and pull it tight against the tacks.

5. Draw an ellipse around the tacks by keeping the string tight against the tacks. This represents Neptune's orbit. Label this ellipse with the word *Neptune*.

6. Move one tack in about one inch and up about one inch.

7. Repeat the process to draw a second ellipse. This ellipse will be larger and will slightly overlap the first. This represents Pluto's orbit. Label this ellipse *Pluto*.

Our Sun

The center of our solar system

LESSON 12

How large is the sun and how hot is it?

Beginners

When you wake up in the morning, you usually see the sun shining brightly. The sun is the center of our solar system. It is a medium-sized star that has a somewhat yellow color. The earth and the other planets in our solar system move around the sun. It takes the earth one year to go completely around the sun one time.

God designed the universe so that the sun would bring light and heat to the earth each day. This is very important because no plants or animals could live on the earth if the sun was not where it is. The sun is just the right distance away to provide enough heat to warm the earth and enough light for plants to grow.

- What is the sun?

- What two things does the sun provide for the earth?

- How long does it take for the earth to go around the sun one time?

The sun is the center of our solar system; but why is it so important to us? The sun provides light and heat to keep us alive. It provides the gravitational pull needed to keep everything in our solar system in its proper place. And the sun was designed by God to rule the day (Genesis 1:14–19).

The sun has a diameter of 868,000 miles (1.4 million km). It is approximately 100 times bigger around than the earth. The sun is so large that one million earths could fit inside it! The sun weighs about 2×10^{27} tons (that's the number two followed by 27 zeroes), which is 333,000 times as much as the earth. 99% of all of the mass of our solar system is in the sun.

Compared to other stars, the sun is a medium-sized star and is yellow in color. The surface temperature is approximately 11,000°F (6,000°C). The sun is composed mostly of helium and hydrogen. Scientists estimate that the sun contains enough hydrogen to continue burning at the present rate for 5 billion more years. As stars continue to burn, the conversion of hydrogen into helium changes the composition of the core, causing it to grow brighter with age. This is another indication that evolution is false. Evolutionists believe that life arose on earth about 3.8 billion years ago. However, the sun would have been 25% dimmer then, and the earth would have been much too cold to support life. This is called the young faint sun paradox, and shows once again that God's Word can be trusted.

Ninety-seven percent of the sun's energy is electromagnetic energy in the form of light, heat, x-rays, and radio waves. It is theorized that the other 3% is in the form of neutrinos, which are believed to be tiny particles that can pass through matter and travel at nearly the speed of light. The energy from the sun moves in waves. These energy waves are visible to our eye if they are between 0.0004 and 0.00075 mm long. Waves that are longer or shorter than this cannot be seen with the human eye. For example, heat waves are longer and x-rays are shorter than visible light.

VIEWING SUNLIGHT

The light coming from the sun has a yellow tint to it. However, Sir Isaac Newton discovered that sunlight actually contains all colors. If the sun emitted all colors equally, it would give off a pure white light. But it emits colors in the middle of the spectrum more strongly so its light appears yellow. A rainbow reveals all of the colors of sunlight. As the light passes through raindrops in the sky, the different colors of light are bent at different angles and are displayed in the rainbow.

Purpose: To create a rainbow and view all the colors of sunlight

Materials: pie pan, small mirror

Procedure:

1. Fill a pie pan with water and place it on a level surface near a window.

2. Place a small mirror in the water so that it reflects onto a wall the sunlight that has passed through the water.

Conclusion:

The light reflected by the part of the mirror that is out of the water will be white, but the light reflected after it has passed through the water will make a rainbow. Water acts like a prism. Different colors of light travel at different speeds through water so the colors spread out and can be seen separately. If you have a prism available, try passing the sunlight through it and see how it separates the colors, too.

All of this information is very interesting, but the most important thing to real-ize about the sun is that God designed it and placed it in just the perfect location for us to live. The sun is just the right distance from the earth to provide heat without burning us up. It provides just the right amount of light for plants to grow and pro-vide the earth with food. The sun may be an ordinary star, but it has an extraordinary purpose for our lives.

For more information on the sun and how it shows God's handiwork, see www.answersingenesis.org/go/sun. ■

WHAT DID WE LEARN?

- What are the main elements found in the sun?
- What colors are found in sunlight?

TAKING IT FURTHER

- Why is the sun so important to us?
- How does energy get from the sun to the earth?

DIAMETER OF THE SUN

Measuring the exact diameter of the sun is a difficult thing. The surface of the sun is constantly changing so exact measurements cannot be made. But scientists have determined that the diameter of the sun is approximately 868,000 miles (1,400,000 km). Using a little math, you can make some simple measurements and calculate the approximate diameter of the sun without needing any fancy scientific equipment. All you need is a pinhole projector.

Purpose: To calculate the diameter of the sun using a pinhole projector

Materials: needle, two index cards, tape, meter stick, ruler, calculator, "Sun Measurement" worksheet

Procedure:

1. Use a needle to punch a small hole in the center of an index card.

2. Tape the index card to the end of a meter (or yard) stick so that the card is perpendicular to the stick.

3. Tape a second card to the other end of the stick so that it is also perpendicular to it.

4. Go outside on a sunny day and place the end of the stick with the card that does not have the hole against the ground.

5. Move the top of the stick until the shadow of the top card falls on the bottom card. You should see a projection of the sun (it looks like a white circle) on the bottom card. **Do not look directly at the sun; it can damage your eyes!**

6. Use a ruler to measure the diameter of the image that is projected onto the bottom card. If you are using a meter stick, measure the diameter of the image to the nearest 1/10 of a millimeter. If you are using a yard stick, measure the diam-eter of the image to the nearest 1/16 of an inch.

7. Record the measurements on the "Sun Measurement" work-sheet. Make at least three mea-surements to try to eliminate error, then find the average of the measurements.

8. Next, follow the instructions on the worksheet to determine the diameter of the sun.

STRUCTURE OF THE SUN

What is it like on the inside?

LESSON 13

What are the different layers of the sun?

Words to know:

chromosphere

corona

sunspot

aurora borealis

aurora australis

core

radiative zone

convective zone

Challenge words:

umbra

penumbra

BEGINNERS

We know that the sun is a star and that it provides heat and light to the earth. Scientists are not completely sure how the sun does this, but they believe that the sun works somewhat like a nuclear reactor. The sun is made from mostly hydrogen and helium. The surface of the sun is extremely hot, over 10,000 degrees Fahrenheit!

You should never look directly at the sun; the sun is so bright that it can damage your eyes. But scientists have special instruments and can take pictures of the sun. They have discovered that some areas on the sun are cooler than others and look darker. These areas are called sunspots. Also, sometimes there are explosions on the surface of the sun. These explosions are called solar flares.

• What are the two things the sun is made from?

• Why should you never look directly at the sun?

• What are cooler areas of the sun called?

The sun is not just a simple ball of hot hydrogen and helium. It has a definite structure that we are just beginning to understand. The sun has an atmosphere that consists of two parts: the chromosphere and the corona. The chromosphere is heated plasma that extends from the surface of the sun to about 6,200 miles (10,000 km) high. The chromosphere is about 11,000°F (6,000°C) at the bottom and up to 1.8 million degrees Fahrenheit (1 million degrees Celsius) at the top. The corona is above the chromosphere. It spreads outward for millions of miles. The corona can reach temperatures of up to 3.6 million degrees Fahrenheit (2 million degrees Celsius). It is continually moving and changing shape.

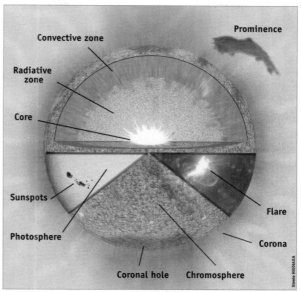

Structure of the sun

The visible surface of the sun is called the photosphere. It consists of plasma—super heated material that is neither solid, liquid nor gas. The surface is covered with granules, which are bubbles of hot plasma rising from the interior of the sun. The photosphere is about 10,800°F (6,000°C), but some areas of the surface are cooler than others. Areas that are only about 8,100°F (4,500°C) are called sunspots. These cooler areas appear to be darker than the hotter areas around them. Sunspots move around the surface of the sun, generally from east to west.

The sun's surface also experiences periods of violent eruptions called solar flares. A solar flare sends out matter, x-rays, and other energy waves that can interfere with radio transmissions on the earth. Solar flares usually last less than one hour. The emissions from these eruptions hit the earth's ionosphere and light it up. This is usually more visible near the poles. In the northern hemisphere this phenomenon is call the aurora borealis, or the northern lights, and in the southern hemisphere it is called the aurora australis, or southern lights.

Solar flare

Scientists cannot directly observe the interior of the sun; however, they have developed a model that they think describes it. They believe that the sun's interior has three parts: the core, the radiative zone, and the convective zone. Scientists believe that the core works like a thermonuclear reactor, generating the sun's energy. They think that hydrogen atoms in the core combine to form helium atoms, which releases huge amounts of energy. This is similar to the reaction that takes place in a hydrogen bomb. It is believed that the temperature in the core of the sun is about 25 million degrees Fahrenheit (14 million degrees Celsius) and that the pressure is 340 billion times earth's air pressure.

Outside the core is the radiative zone. Energy moves outward from the core through the radiative zone as electromagnetic waves. As the energy approaches the

FUN FACT If the scientists' model of the sun is correct, the sun loses about 6 million tons of material per second. Five million tons are converted into energy and 1 million tons are blown away in the solar wind. Because the sun uses up so much material, it is slowly wearing out. No evolutionary theory adequately explains what gave the stars their energy to begin with. Only God could have started the billions of power plants (stars) we see throughout the universe.

surface it passes through the **convective zone** where the hot gas cools slightly, falls back to the radiative zone, gets heated again, rises to the surface, and so on. This is why the surface of the sun resembles a pot of boiling water (see photo at right).

As scientists continue to study the sun, they will better understand its makeup. ■

WHAT DID WE LEARN?

- What are the two parts of the sun's atmosphere?
- What is a sunspot?
- Are sunspots stationary?
- What do scientists believe are the three parts of the sun's interior?

TAKING IT FURTHER

- What is the hottest part of the sun?
- What causes the aurora borealis or northern lights?
- When do you think scientists study the sun's corona?

TRACKING THE EARTH'S MOVEMENT

The earth rotates with respect to the sun. It makes one complete rotation every 24 hours. We can see this relative movement as we watch the sun move across the sky throughout the day. Another fun way to track the movement of the earth is to watch your shadow move.

Purpose: To track the movement of the sun

Materials: sidewalk chalk

Procedure:

1. On a sunny day, go outside early in the morning and make an X on the sidewalk or driveway, then stand on the X.
2. Have someone trace your shadow on the ground using sidewalk chalk.
3. Write the time next to the shadow.
4. Repeat this activity every 2–3 hours throughout the day.

Conclusion:

You will see your shadow get shorter and shorter until the sun reaches its highest point in the sky. Then the shadow will begin to grow longer in the other direction as the sun begins to set.

SUNSPOTS

Sunspots are areas that are cooler than the surrounding surface of the sun. However, scientists do not fully understand what causes sunspots. They know that sunspots are associated with areas of higher magnetic fields. The magnetic field in a sunspot is about 1,000 times greater than in the rest of the sun. It is believed that this greater magnetic field somehow hinders the convection of heat from the interior of the sun, causing the sunspots to be cooler than the surrounding areas.

The center of a sunspot is called the **umbra** and is the darkest area. The outer edge of a sunspot is called the **penumbra** and is warmer and brighter than the umbra, but not as hot as the rest of the surface of the sun (see close-up at right).

Sunspots rotate with the rotation of the sun. Because the sun is not a solid like the earth, but plasma, all of the mass of the sun does not rotate at the same speed. The mass at the equator rotates about once every 25 days but the mass near the poles rotates about once every 35 days. Thus the sunspots near the equator move across the surface faster than the sunspots that occur at higher or lower latitudes.

Some sunspots last for a few days, and other sunspots last for several weeks. Records of sunspots over the past 150 years have shown that sunspots occur in cycles. Some years there are relatively few sunspots and other years many sunspots occur. This pattern repeats approximately every 11 years. Years when there are few sunspots are called solar minimum, and when the number of sunspots peaks it is called solar maximum.

Solar flares appear to be related to sunspots because they almost always occur near sunspots and there are more solar flares when there are more sunspots; however, the exact cause of solar flares is not completely understood and therefore solar flares cannot be predicted.

You may be able to detect sunspots using your pinhole projector. Look at the projected image of the sun and see if you can detect any dark spots in the image. Your image will be very small so you may not be able to see them.

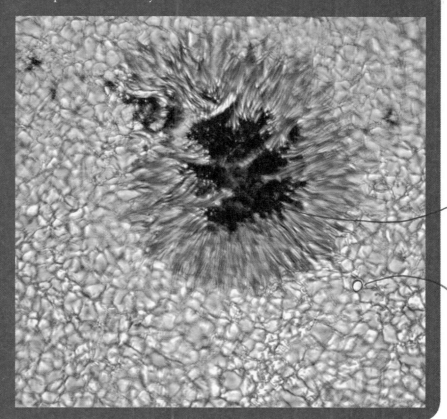

Everyone Else

my skin tone

SOLAR ECLIPSE

Where did it go?

LESSON 14

What causes an eclipse?

Words to know:

solar eclipse

lunar eclipse

BEGINNERS

The earth is moving around the sun, and the moon is moving around the earth. Sometimes the moon comes directly between the earth and the sun. If the moon is in just the right place, it can block the light from the sun. When this happens we call it a solar eclipse. A solar eclipse is visible from somewhere on earth from one to three times per year.

When an eclipse occurs, it can completely block the sunlight in an area on earth about 150 miles across and block some of the sunlight in an area that is much bigger, but the eclipse never blocks out all of the light to the whole earth. You have to be in just the right location to observe a solar eclipse when it happens.

If you happen to be in the right place at the right time to observe a solar eclipse, you should not look directly at the sun. Although the moon blocks some or even most of the light from the sun, it does not block all of the harmful radiation and you can still damage your eyes. The time the moon first moves in front of the sun until it moves completely across the sun can be about 2 hours, but the time that the sun is totally blocked during an eclipse is usually less than 7 minutes.

- **What is a solar eclipse?**

- **Does an eclipse block the light from the whole earth at once?**

- **How often does an eclipse happen?**

An eclipse occurs when one heavenly body blocks the light from another heavenly body. From the earth's point of view, there are two types of eclipses. A solar eclipse is when the moon blocks the sun's light and casts a shadow on the earth. A lunar eclipse is when the earth blocks the sun's light and casts a shadow on the moon. We will study more about lunar eclipses in lesson 17.

A solar eclipse occurs when the moon comes directly between the sun and the earth. This can only happen when the moon is at its new moon phase. A solar eclipse happens somewhere on the earth one to three times each year. The moon's orbit around the earth is at a 5-degree tilt to the earth's orbit around the sun, so it usually goes above or below the plane of the earth's orbit.

When an eclipse does occur, it can be either a partial or a total eclipse. During a partial eclipse, the moon covers part of the sun's disk but not all of it. It is much more likely that a person will observe a partial eclipse than a total eclipse. Partial eclipses happen more frequently and can be observed in a much larger area.

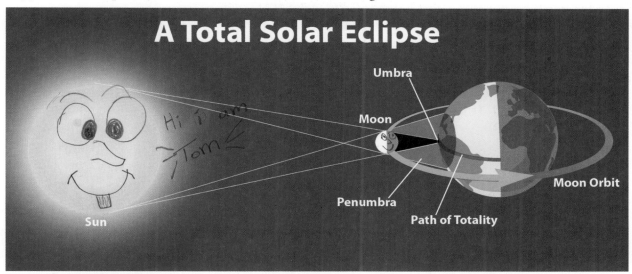

A Total Solar Eclipse

Umbra

Moon

Sun

Penumbra

Path of Totality

Moon Orbit

A total eclipse occurs when the moon covers the entire disk of the sun. Only the corona and chromosphere of the sun can be observed during the total eclipse. A total eclipse can be observed in only a small area of the earth, usually an area about 150 miles (240 km) in diameter. During a total eclipse, there does not appear to be much change in the amount of sunlight until the sun is almost completely covered. Then the sky darkens and plants and animals begin to act as though night was falling. At

FUN FACT

There is no place in the universe quite like earth. Only on earth do the moon and sun appear to be about the same size in the sky. Although the moon is about 400 times smaller than the sun, it is also about 400 times closer to the earth. This allows the earth to experience solar eclipses. These eclipses can be accurately predicted and have been used to date events in the past. God has truly provided the sun and the moon to mark the seasons, days, and years just as He said in Genesis 1:14–15.

MAKING AN ECLIPSE

Purpose: To simulate a solar eclipse

Materials: large ball, flashlight, small ball

Procedure:

1. Have one person hold a large ball such as a basketball or volleyball. This represents the earth.

2. Have another person stand 3 to 4 feet away and shine a flash-light on the ball. This represents the sun.

3. One of you hold a small ball such as a tennis ball in the middle of the light beam. This represents the moon.

4. Have the person with the "moon" move it until it blocks out the light from the flash-light, casting a shadow on the "earth."

Conclusion:

Notice how much of the earth is covered by the shadow. It will be a small percentage. Experiment with the moon in different locations. Watch what effect this has on the light reaching the earth. You will find that the moon must be in a very specific place in order to cause an eclipse.

totality, the sky is as dark as a moonlit night. Totality can last for as little as an instant to as long as 7½ minutes. The total time for the moon to move across the face of the sun is usually about two hours.

It is vitally important to remember that even if the moon covers the sun, there is still a significant amount of dangerous radiation coming from the sun. So, never look at the sun, even during an eclipse! An eclipse should be viewed by projecting an image of the sun on a box through a hole in one end of the box and watching the shadow. To learn more about eclipses, go to the NASA web site at http://sunearth.gsfc.nasa.gov/eclipse. ■

WHAT DID WE LEARN?

- What is an eclipse?
- What is the difference between a partial and a total eclipse?
- How often do solar eclipses occur?

TAKING IT FURTHER

- Why do you think plants and animals start preparing for nightfall during an eclipse?
- Why can a total eclipse only be seen in a small area on the earth?
- How can the moon block out the entire sun when the sun is 400 times bigger than the moon?

Sun & Moon

TOTAL SOLAR ECLIPSE

The next total solar eclipse visible in the United States will occur on August 21, 2017. It will be spectacularly visible to millions of people across the United States. This will be an impressive total eclipse, lasting over 2½ minutes at maximum and visible over a path up to 71 miles (115 km) wide.

The path will start in the Pacific, well north of Hawaii, and then cross to make landfall in the U.S. in the northern half of Oregon. It then will cross Idaho, Wyoming, and Nebraska, and clip the northeast corner of Kansas before passing right over Missouri. It will also cut over the southern tip of Illinois and the western end of Kentucky, before crossing Tennessee and the western tip of North Carolina. The extreme northeast corner of Georgia will also be in the path of totality.

Research past and future solar and lunar eclipses. Make a chart showing when and where the eclipses have and/or will occur. You can find information on eclipse schedules in several places. You may wish to start with the NASA Eclipse Home Page, http://sunearth.gsfc.nasa.gov/eclipse/eclipse.html. You are likely to live in an area where you can see a lunar eclipse, but you are less likely to live in an area where you can view a total solar eclipse.

Sun & Moon

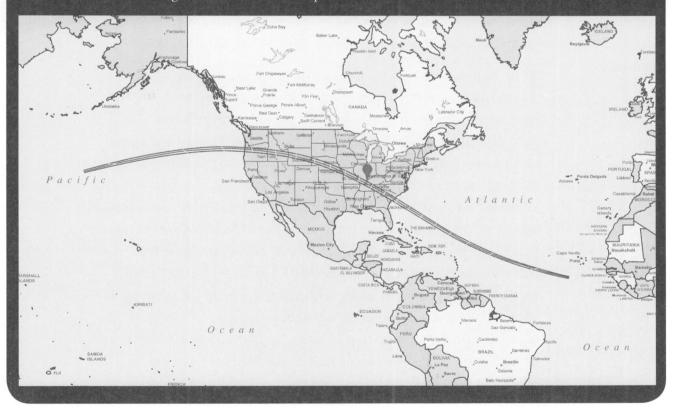

Solar Energy

Can it meet our energy needs?

LESSON 15

How can we use the sun for energy?

Words to know:

solar energy

Beginners

The heat and light from the sun help us in many ways. The heat warms the earth so that plants and animals can live here. The light provides light for people and animals during the day. The light also helps plants to grow and make food. People are trying to find new ways to use the heat and light from the sun to provide other kinds of energy.

Solar energy is energy we get from the sun. There are two main ways that people are using the energy from the sun. One way is to build solar panels or solar collectors that collect the heat from the sun and use it to heat water. You may have seen solar collectors on the roof of a house. The second way that people are using solar energy is with special solar cells that convert the light directly into electricity. You may have used a solar calculator that uses the light from the sun to power the calculator. Scientists are working on other ways to use the energy from the sun to help provide power for people's lives.

- What is solar energy?

- What are two ways that people use solar energy?

The sun is a giant power plant. The energy that reaches the earth from the sun heats the globe and helps plants to grow. The amount of energy that reaches the earth from the sun in only two weeks' time is equal to the energy in all of the coal, oil, and natural gas reserves in the world. Fossil fuels such as coal, oil, and natural gas cannot be easily replaced and may eventually be used up. Therefore, many people are looking to solar energy—the energy from the sun—as a possible replacement. Scientists are developing ways to collect and use this huge renewable source of energy.

Solar energy has many advantages over fossil fuels. It is cleaner—it does not produce pollution such as smoke or nuclear radiation. It is also renewable. In fact, for all intents and purposes it is infinite. And it is a high-quality source of energy that is easily converted into heat. But there are disadvantages as well. Solar energy is not always available. It is not available at night, and in areas near the poles it is not available at all during much of the winter. Also, solar energy is very dispersed or spread out. It must be concentrated in order to be used.

For years, gardeners have used solar energy to heat their greenhouses. This is the idea behind solar collectors. Solar collectors are essentially flat boxes that have been painted black on the inside. Tubes carrying water are placed inside the boxes and then the boxes are covered with glass. As the sun passes through the glass, some of the energy is trapped. This energy heats the water as it flows through the pipes. Homes with solar panels have a nearly free supply of hot water. Also, this hot water can be used to heat the homes on cold days.

In the past few decades, many researchers around the world have been working on ways to collect and concentrate the rays of the sun so that they can be used to generate electricity, not just heat water. A solar power plant may have an array of hundreds of solar collectors that concentrate the sun's energy and heat water that turns to steam and drives the turbines to generate electricity. But there are relatively few solar power plants in operation today.

A solar cell works differently than a solar collector. A solar cell takes the sun's light energy and converts it directly into electricity. These cells are called photoelectric or photovoltaic cells. Solar cells are wafers made from cadmium sulfide,

Many houses have solar panels on the roof to generate electricity.

FUN FACT

Researchers have built a solar furnace in France. This facility has thousands of flat mirrors that reflect the sun's light toward a ten story curved mirror. The curved mirror concentrates all of the sunlight at a tower. This concentrated sunlight can raise temperatures inside the tower to over 5,400°F (3,000°C). This facility is operated by the Solar Research Institute.

SOLAR ENERGY

Activity 1—Purpose: To test which color absorbs the most heat

Materials: black paper, green paper, white paper, baking sheet, ice cubes, "Solar Energy" worksheet

Procedure:

1. Place a piece of black paper, a piece of green paper, and a piece of white paper side by side on a baking sheet.

2. Place an ice cube on each sheet of paper.

3. Set the baking sheet in a sunny location.

4. On the "Solar Energy" worksheet, record the time you set the sheet in the sun as the starting time and the time the ice was completely melted as the ending time.

5. Answer the questions on the worksheet.

Activity 2—Purpose: To demonstrate how the sun heats water in a solar collector

Materials: two clear glasses, black paper, tape, two thermometers, "Solar Energy" worksheet

Procedure:

1. Fill two clear glasses with water.

2. Wrap black paper around one glass and tape it in place.

3. Place both glasses side by side in a sunny location.

4. On the "Solar Energy" worksheet, write down the starting time of your experiment.

5. Use a thermometer to measure the temperature of the water in each glass and record it on the worksheet.

6. Repeat the temperature measurement every 5 minutes for 20 minutes and record the results on the worksheet.

7. Answer the questions on the worksheet.

---

silicon, or gallium arsenide. Solar cells have two layers of material. Each layer has a different electrical charge. When the sunlight hits the top layer, the charges begin to flow from one layer to the other, thus creating a current of electricity. Solar cells can convert about 15% of the light that hits them into electricity. Researchers are trying to find ways to increase this percentage.

Solar cells work on cloudy days just as well as they do on sunny days; however, they do not work at night. Solar cells are used in many places. They are used to power street signs and lights, especially in remote areas where there is not a readily-available source of electricity. They are also used as back-up energy sources in many areas. Chances are, you have seen a solar operated calculator that works off of the sun's light or even the light in your room. But the biggest use of solar cells is in space technology. Solar energy is abundant in space where there is no atmosphere to block the light. Solar cells are used to generate electricity for satellites, space probes, and even the space station.

More research is needed before solar energy can be used to completely replace our current sources of energy. But God has supplied us with a wonderful source of energy, and we will continue to find ways to use it. ■

Sun & Moon

WHAT DID WE LEARN?

- What is solar energy?
- What are the two ways that solar energy is used today?

TAKING IT FURTHER

- Why is solar energy a good alternative to fossil fuels?
- Why are the insides of solar collectors painted black?
- What are some of the advantages of using solar cells in outer space?

SOLAR ENERGY

One of the problems with using solar energy efficiently is that the light from the sun is dispersed, or not very concentrated. Because the sun is far away, the light spreads out before it reaches the earth. This is good for plants and animals on the earth, but it presents some challenges for engineers trying to harness this energy. The light is not evenly distributed on the earth either. Because the earth is tilted at a 23-degree angle with respect to the sun, the rays from the sun hit the hemisphere that is tilted toward it more directly than the hemisphere that is tilted away from it. This is why summers are warmer than winters.

Purpose: To demonstrate the distribution of light on earth

Materials: hard back book, flashlight, clipboard, piece of paper, pencil

Procedure:

1. Place a hard back book upright on a table with the cover spread slightly so it will stand up.

2. Place a flashlight on top of the book and point the flashlight at a clipboard with a piece of white paper on it. Be sure that the clipboard is vertical.

3. Trace the circle of light made by the flashlight on the paper.

4. Now, tip the clipboard away from the flashlight so the light is not hitting it as directly and trace the pattern of light made by the flashlight on the paper.

Questions:

- Is the new pattern bigger or smaller than the first pattern?
- Based on what you just learned, where would be the best location for a solar energy power plant?

OUR MOON

Is it made of green cheese?

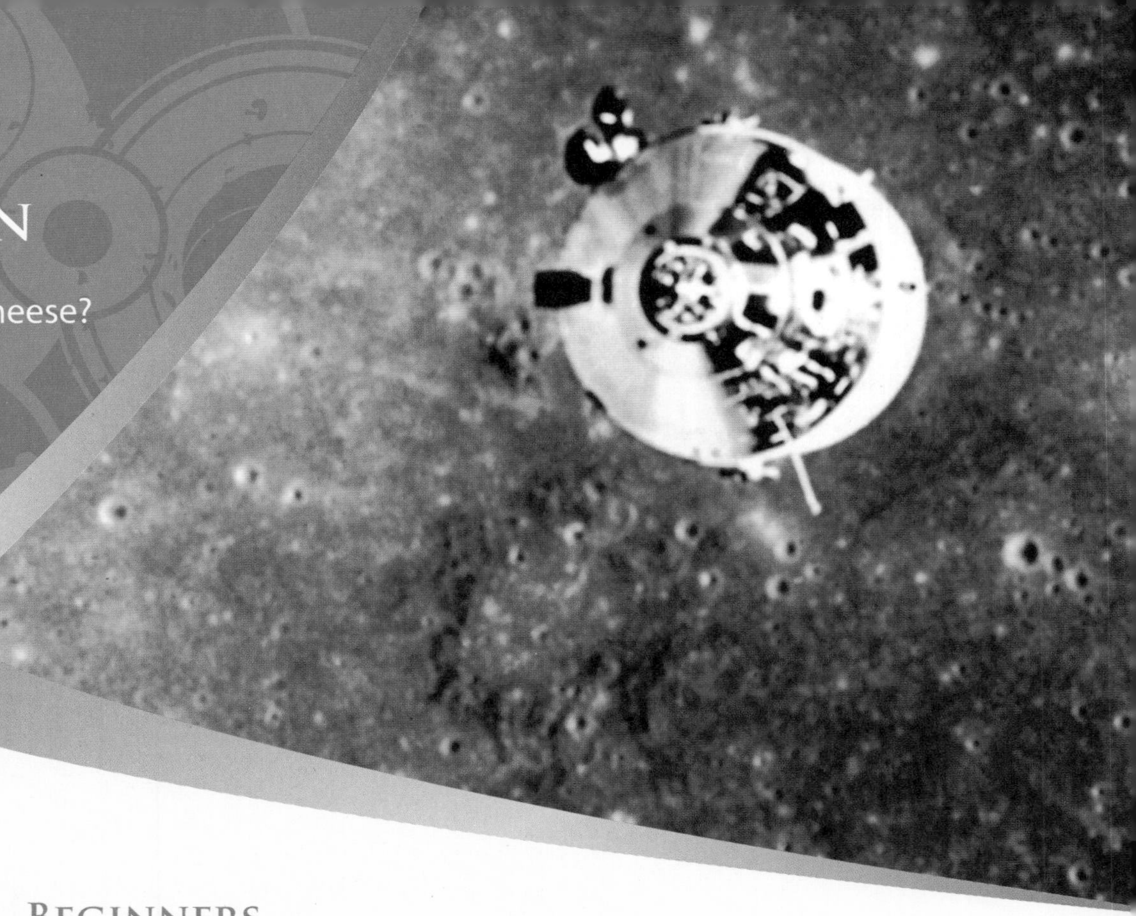

LESSON 16

What is the moon made of and what is its surface like?

Words to know:

maria

Challenge words:

ray

rill

BEGINNERS

Have you ever been outside on a night when there was a full moon? The light reflected from the moon makes it easier to see things at night. The moon is a large round rock that orbits the earth about once each month. The moon appears to shine on the earth, but it does not actually produce any light of its own. The moon reflects the light of the sun toward the earth so that there is light at night as well as in the daytime.

The moon is much smaller than the earth and has less gravity. This is why astronauts can leap long distances on the moon. The surface of the moon is very different from the surface of the earth. There is no air to breathe so there are no living plants or animals. There are just rocks, hills, valleys, and craters on the moon. God did not design the moon for us to live there, but He did design it to give us light.

• Does the moon make its own light?

• Where does the moon's light come from?

• How is the moon's surface different from the earth's surface?

The Bible tells us that God created the sun to rule the day and the moon to rule the night (Genesis 1:16). Although the moon lights up the night, it does not generate any light itself. Instead, it reflects light from the sun. Without the moon, nighttime would be very dark indeed.

The moon is a sphere of rock that orbits the earth once every 27.3 days, or about once per month. The word *month* comes from the old word for moon. The average distance from the center of the earth to the center of the moon is about 239,000 miles (384,500 km). The moon's diameter is 2,159 miles (3,475 km) across, less than the width of the United States. The moon is 1/81 as massive as the earth and has 1/6 the gravity. Because of the low gravitational pull, the moon does not have an atmosphere. Without an atmosphere, there is no protection from the dangers of space, including extreme temperature swings and meteorites. This is why the surface of the moon is covered with craters. There are hundreds of thousands of craters ranging in size from a few inches to hundreds of miles across. A few of the craters do not appear to have been formed by impact. Some scientists think these may have been formed by past volcanic activity on the moon.

In addition to craters, the surface of the moon also has many dark areas called maria (MAH-ree-uh). This is the Latin word for seas. Early astronomers thought that these areas were bodies of water. These areas are actually broad plains that are covered with hardened lava that is darker than the surrounding area. The maria on the near side of the moon cover nearly half of the surface. Most maria are believed to be impact craters that later filled with lava. The far side of the moon also has many craters, but most of them are not filled with lava. In addition to craters, the moon's surface also has many mountain ranges and valleys.

The moon that orbits earth is unique among all the moons in our solar system. It is 100 times larger than the average moon. This allows it to reflect a substantial amount of sunlight to the earth, making the night more pleasant. God truly provides for all of our needs. ■

FUN FACT

Some people say the maria on the moon resemble a face, and thus they call the picture made by the maria "the man in the moon." Also, the pock-marked appearance of the moon has given it the reputation of being made of cheese, which, of course, is not true.

WHAT DID WE LEARN?

- Why does the moon shine?
- What causes the dark spots on the surface of the moon?

TAKING IT FURTHER

- Why does the size of our moon show God's provision for man?
- Why is gravity much less on the moon than on the earth?
- Why doesn't the surface of the earth have as many craters as the surface of the moon?

REFLECTED LIGHT

Purpose: To demonstrate how the moon reflects the sun's light

Materials: reflector, flashlight

Procedure:

1. Stand in a dark room holding a reflector. Observe how much light it gives off. (It should be none.)

2. Next, have someone stand across the room and shine a flashlight at the reflector. Notice how much light the reflector now gives off.

3. Experiment with different angles and positions of the reflector to the flashlight. See when the light from the reflector is the brightest and when it is the dimmest. Is there an angle at which the reflector gives off no light?

Conclusion: Remember that the reflector does not produce any light, but rather redirects the light of the flashlight. Just as the reflector reflects the light of the flashlight to our eyes, so the moon reflects the light of the sun to the earth. For more information on the moon and how it shows God's handiwork, see www.answersingenesis.org/go/moon.

SURFACE OF THE MOON

The surface of the moon is marked primarily by craters, maria, mountain ranges, and rills. The craters are indentations in the surface, usually with high sides. It is believed that most of the craters were formed when meteorites struck the surface of the moon. Many times meteorites strike with such force that they explode and sometimes even melt the surface of the moon where they land. Radiating out from many of the craters are bright streaks called **rays**. These rays are believed to be lunar material that was blown outward by the explosion of the meteorite. During a full moon, rays can be seen stretching out from many of the craters. Some rays stretch out as far as 1,000 miles (1,600 km).

In addition to the many craters on the moon, there are many areas called maria, which as you read earlier in the lesson are plains covered in lava. Because the maria resembled seas, they were given names that reflected weather conditions including the Sea of Rains, the Sea of Tranquility, the Sea of Storms, and the Sea of Serenity. When people look at the moon, they often claim to see pictures formed from these dark areas. The most famous of these "pictures" is the man in the moon.

Rising from the surface of the moon are also many mountain ranges. These mountains appear to be similar to mountains on earth minus the vegetation. Some mountains are nearly 5 miles (8 km) high, which is similar in height to Mount Everest and the other very high peaks on earth. Many of the mountain ranges on the moon were named after the mountain ranges on earth including the Alps, Caucasus, and Carpathian Mountains.

Finally, the surface of the moon also has many valleys that are called **rills**. Some rills look like winding rivers, but do not carry water. It is believed that these rills were probably paths for the lava that flowed in the maria. Other rills are fairly straight and resemble fault lines on earth.

Using a pair of binoculars or a telescope, observe the moon and see how many of the features mentioned in this lesson you can observe. Also, look at the maria and see what pictures you can see on the face of the moon.

Newton & the Apple

1642–1727

Isaac Newton led an interesting life, and according to the calendar used in England at the time, he was born on Christmas Day, 1642. However, some people would disagree on his birth date because the calendar in use in England at that time was off by 10 days from the calendar in use today. By our calendar, he was born on January 4, 1643. But this did not matter to the people of that day for it was Christmas to them.

Isaac never knew his father, because he died three months before Isaac was born. His mother remarried when Isaac was two and sent him to live with his grandparents. When Isaac was about ten, his stepfather died and Isaac moved in with his mother, stepbrother, and two stepsisters for a time. He was later sent away to school and lived with a family named Clark. At this time he showed little promise in school and the school reports said he was, "inattentive and idle," so he returned home.

His mother, now a woman of reasonable wealth and land, thought Isaac was now old enough to manage her affairs. He soon showed this was not the right job for him. After this failed endeavor, Isaac's uncle persuaded his mother to let him try school again. This time he stayed with Stokes, the headmaster of the school, and he did much better. He did well enough that his uncle persuaded his mother to let him go to the university.

Although Isaac's mother was fairly well off, when Isaac entered Trinity College at Cam-

bridge in 1661, he entered as a sizar, one who works as a servant to other students. By doing this, he received an allowance toward his college expenses. It's interesting to note that at 18 he was older than most boys entering the college.

When he started at Cambridge, he was planning to receive a law degree, but sometime in college he found a whole new respect for mathematics. He was elected a scholar and received his bachelor's degree in 1665. It was in that summer that the college was closed for two years because of an outbreak of the plague.

When the school closed, Isaac returned home and it was there he started revolutionizing astronomy, optics, physics, and mathematics. He performed many experiments and developed many mathematical formulas while waiting for the plague to pass. When the school reopened after two years, Isaac returned and earned his masters degree. It was after he returned to school that Isaac Newton began some of his most famous work.

One of Isaac's most important discoveries was that sunlight was not just simple white light but was made up of all the colors of the rainbow. He proved this by passing sunlight through a glass prism and producing a rainbow. This discovery led him to develop a telescope using mirrors instead of just lenses in order to eliminate the distorted colors caused by the lenses.

Even though Isaac Newton was a very bright man, he was not able to handle criticism very well. Because of the criticism he received from one of the other scientists in the field of optics, Newton quit the Royal Society and did not publish his findings on optics until after his critic died, some 30 years later.

This did not stop Newton from continuing his research in other areas, however. Isaac Newton is most well known for his work in physics. He was the first to come up with the idea that the earth's gravity influences the moon. He is reported to have been sitting in a garden when he observed an apple fall to the ground. This gave him the idea that the force working on the apple might also be the force that keeps the moon revolving around the earth. After much research and testing, Newton devised the law of gravitation and was able to explain the gravitational pull mathematically. The law of gravitation states: "Any two bodies attract each other with a force proportional to the product of their masses and inversely proportional to the square of the distance between them." This means that heavier objects exert more gravitational pull than lighter objects and that the gravitational pull decreases the farther the two objects are from each other. This discovery helped him explain the tides and the orbit of comets, as well as the orbit of the moon.

Newton made many other advancements in the areas of science and mathematics. From 1669–1687, while he was a professor at Cambridge, he did most of his productive research. Newton is credited with laying the foundations for calculus and detailed his work on physics in his book *Philosophiae Naturalis Principia Mathematica*, or *Principia* as is it better known. *Principia* is considered by many to be the greatest scientific book ever written. In fact, Newton made so many discoveries that in 1705 he was knighted by Queen Anne. He was the first scientist so honored for his work.

Sir Isaac Newton had a deep faith in God and believed in the Bible's account of creation. In his works he expressed a strong sense of God in all of nature. In *Principia* he wrote, "This most beautiful system of the sun, planets, and comets, could only proceed from the counsel and dominion of an intelligent Being. ... This Being governs all things, not as the soul of the world, but as Lord over all; and on account of his dominion he is wont to be called "Lord God" ... or "Universal Ruler". ... The Supreme God is a Being eternal, infinite, absolutely perfect."

Although we think of Newton as a great scientist, he did not spend his entire life on research. In 1689 Newton was elected as a member of parliament representing the University and was reelected in 1701. In 1696, Isaac Newton was given a position in the government as Warden of the Royal Mint. Then three years later, he was named Master of the Mint. He was able to lead the mint through a difficult time of recoinage, and he worked to eliminate counterfeiting. In 1703 he was elected President of the Royal Society and kept that position until his death in 1727. Newton spent only 20 years of his career on science and math research, yet the discoveries made by Sir Isaac Newton helped shape science as we know it today.

Motion & Phases of the Moon

There's a full moon tonight

LESSON 17

How does the moon move and why does its appearance change in our sky?

Words to know:

new moon

waxing crescent

waxing gibbous

full moon

waning gibbous

waning crescent

Challenge words:

near side of the moon

far side of the moon

light side of the moon

dark side of the moon

BEGINNERS

The moon spins or rotates just like the earth does. It also revolves around the earth. The moon rotates and revolves at about the same rate, once per month, so the same side of the moon is always facing the earth.

Because the moon is always moving around the earth, each day it is in a slightly different place compared to the sun, so the light from the sun reflects off of it in a different way. This is why sometimes we have a full moon and sometimes we can only see a small part of the moon shining in the sky. When no light reflects off of the moon, it is called a **new moon**. The next night a little light will reflect off the moon. Then the amount of light will grow each night until there is a **full moon**. Then the amount of light will go down each night until there is another new moon. The time between new moons is about one month.

- **What two ways does the moon move?**

- **What is a new moon?**

Just as the earth moves in two different ways with respect to the sun, so also the moon moves in two ways with respect to the earth. The moon revolves around the earth, and it rotates on its axis. From the perspective of the earth, the moon revolves around, or circles, the earth about once per month. The moon rotates on its axis once per month as well. This results in the same side of the moon always facing the earth. This side is called the near side of the moon. The back side is called the far side of the moon.

Because the moon orbits the earth, it is in a different position with respect to the sun each day for about a month, and then it repeats the cycle. The moon's position with respect to the sun determines how much light reflects off of its surface to the earth. Therefore, the moon appears to change shape throughout the month. These varying shapes are called the phases of the moon.

At new moon, the moon is between the earth and the sun. We can't see the moon at all because the sun is too bright and shining only on the far side of the moon. A couple nights later, we can see a small sliver lit up on the right side of the moon. This is called a waxing crescent. After seven days, the right half will be lit up. This is called the first quarter. As the moon continues to "grow" throughout the next week it is called a waxing gibbous. At 14 days through the cycle, the moon is at the opposite side of the earth from the sun and is considered a full moon. The full moon rises at sunset, and sets at sunrise. A full moon is nine times brighter than the moon when it is half lit (during the first or last quarter). The line between the light and dark sides of the moon is called the moon's terminator.

The next day, after the full moon, the moon appears to get smaller. This is called a waning gibbous. Then ¾ of the way through the cycle, only the left half of the moon is lit and is called the last quarter. Finally, as the moon "shrinks" further it is called a waning crescent.

Even though it takes 27.3 days to complete one orbit, the moon's cycle from one new moon to the next takes 29.5 days, or one synodic month, to complete. If you viewed the earth and moon from a location in space, you would see the moon line up with a particular point on the earth every 27.3 days. But the moon lines up between the earth and the sun the same way every 29.5 days because they are both moving around the sun as the moon moves around the earth.

Just as the moon passing directly between the earth and the sun causes a solar eclipse, so too, if the earth passes directly between the sun and the moon we see a lunar eclipse. This can only occur at full moon and only when the sun, earth, and moon are all directly lined up. This happens infrequently because the moon's orbit is tilted with respect to the earth's orbit. During a lunar eclipse, the moon is still visible with a slight red color. The longest lunar eclipse was recorded at 1 hour 40 minutes of complete eclipse with the moon being at least partially blocked for 3 hours and 40 minutes. ■

MOON PHASES

Label the moon on the "Identifying Phases of the Moon" worksheet.

WHAT DID WE LEARN?

- What causes the phases of the moon?
- Why does the same side of the moon always face the earth?
- What causes a lunar eclipse?
- From the perspective of space, how long does it take for the moon to complete its cycle around the earth?

TAKING IT FURTHER

- Why doesn't a lunar eclipse occur every month?
- What is the difference between a waxing crescent and a waning crescent?

OBSERVING THE MOON

There are some terms associated with the moon that can be confusing. Since the moon rotates and revolves at the same rate, the same side of the moon is always facing the earth. As we mentioned before, this is called the **near side of the moon**, and the back side is called the **far side of the moon**.

You may also hear the terms light and dark sides of the moon. You may think that since the moon reflects the sun's light toward the earth that the side facing the earth would be the light side of the moon, but this is not necessarily true. The **light side of the moon** is the side facing the sun, regardless

of the moon's position with respect to the earth. And the **dark side of the moon** is the side facing away from the sun. When is the light side of the moon the same as the near side of the moon? When is the dark side of the moon the same as the near side of the moon?

Because there is no atmosphere

on the moon, the temperatures on the moon vary greatly. On the light side, when the sun is shining directly on the moon, the temperature can be as much as 265°F (130°C) and on the dark side of the moon the temperature can be as low as −280°F (−173°C).

Phase	Moon Rise	Moon Set
New Moon	6 A.M.	6 P.M.
First Quarter	Noon	Midnight
Full	6 P.M.	6 A.M.
Third Quarter	Midnight	Noon

Purpose: To observe and record the phases of the moon

Materials: "Observing the Phases of the Moon" worksheet

Procedure:

1. On a copy of "Observing the Phases of the Moon" worksheet, fill in today's date under the first circle, tomorrow's date under the second circle, and so on for a total of 29 days.

2. Next, fill in the time of the moon rise and moon set for each date. You can often find the moon rise and set times in your local paper or on the internet. Above is a chart that shows approximate rise and set times for each phase of the moon, but local times will vary.

3. Now that you have your chart ready, observe the moon each day for the next month. Use a pencil to fill in the part of the moon that appears dark. This will give you your own chart of the phases of the moon.

Conclusion:

After observing the moon every day for one month, you will have a better understanding of how the moon moves through the sky and how its position changes each day with respect to the sun.

Sun & Moon

ORIGIN OF THE MOON

Where did it come from?

Where did the moon come from?

BEGINNERS

You may wonder where the moon came from. Was it always orbiting the earth, or was there a time when the moon was not there? The Bible gives us the answer to these questions. According to Genesis 1:1–19, God created the earth on the first day of creation and He created the sun, moon, and stars on the fourth day of creation. He spoke them into existence from nothing. Aren't you glad God made the earth, sun, moon, and stars?

Some people do not believe in God and have tried to explain where the moon came from without accepting creation. One of these ideas is called the Capture Theory. This idea says that the moon was originally going around the sun, but something knocked it out of its orbit and it was captured by the earth's gravity and began to orbit the earth instead. There is no explanation for how the moon was knocked out of its original orbit. Also, the chances that it would come close to the earth at just the right speed and from just the right direction to be captured is nearly zero. Believing in God provides a much better answer to where the moon came from.

- Where did the earth come from?
- Where did the moon come from?
- On which day of creation did God create the moon?

Genesis says that God created the sun, moon, and stars on Day Four of creation. Yet many scientists do not accept the biblical account of creation. Evolutionists have proposed many theories to explain where the moon came from, and how it ended up orbiting the earth. However, all of these theories have significant problems.

One theory, called the Capture Theory, says the moon originally orbited the sun. Then, somehow it was dislodged from its orbit and was captured by the earth's gravity. There are no observed or credible ideas for what could have caused the moon's original orbit to be disrupted. Also, the chances that the moon would have approached the earth at just the right speed and angle to be captured are extremely slim. It is much more likely that the moon would have flown off or been pulled into the earth than to begin orbiting it.

A second theory, called the Fission Theory, says that as the earth was spinning, it spun a ball off of itself that became the moon. However, in order for the ball to escape the gravity of the earth and begin orbiting it, the earth would have had to be spinning at about one revolution every 2–3 hours. Again, this would create tremendous heat. There is no evidence of such heat in the geologic records. Also, the composition of moon rocks is different from that of most earth rocks. If the moon came from the earth, the moon's rocks should have the same composition as the earth's rocks. Most scientists have disregarded the Fission Theory because of its problems.

A third idea, called Accretion, says that the earth, moon, and other planets were all formed when particles in a giant dust cloud generated by the big bang started sticking together. Scientists cannot explain what would cause these particles to stick together. Gravity of small particles is very small, so they would not be attracted to each other by gravity. Also, the average density of the earth is much greater than the average density of the moon. If these two bodies were formed from the same

SPINNING BODIES

Purpose: To demonstrate why the capture theory is wrong

Materials: two tops, masking tape

Procedure:

1. Place a piece of masking tape on the floor. Write the letter *E* on it to represent the orbit of the earth.

2. Place a second piece of tape about three inches away and write an *M* on it to represent the moon's orbit.

3. Spin the "earth" top on the floor on the *E* piece of tape.

4. Then, start the "moon" top spinning, but not on the *M*. Try to make the "moon" top hit the "earth" top in such a way that the moon top ends up spinning on the *M* after the collision without significantly moving the "earth" top from the *E*.

5. Repeat this several times.

6. Now, start each top spinning on its designated spot.

Conclusion:

You will find that it is impossible, or nearly so, to get the tops to end up on the right pieces of tape. The actual science involved is much more complicated than this simple demonstration, yet this is how the Capture Theory tries to explain the origin of the moon. Step 6 demonstrates how God created the earth and the moon—He put them in their places.

FUN FACT

The moon is slowly moving away from the earth (called recession). As the moon orbits the earth, its gravity pulls on the earth's oceans, causing tides. The tides actually "pull forward" on the moon, which causes the moon to gradually spiral outward. Today, the moon moves about an inch and a half away from the earth each year, which means that the moon would have been closer to the earth in the past. This rate of recession would have been greater in the past when the moon was closer to the earth.

Six thousand years ago, the moon would have been about 800 feet (245 m) closer to the earth (which is not much of a change, considering the moon is a quarter of a million miles away). So this recession of the moon is not a problem over the biblical timescale of 6,000 years. But, if the earth and moon were over four billion years old (as evolutionists teach), then we'd have big problems. In this case, the moon would have been touching the earth only 1.4 billion years ago, suggesting that the moon can't possibly be as old as secular astronomers claim.

gas cloud, why do they have such different densities? No one can adequately answer this question.

A fourth theory, the Impact Theory, states that something the size of Mars hit the edge of the earth, knocking off a chunk, which became the moon. This is extremely unlikely. There is no evidence that this has happened. Collisions with objects as large as planets have never been observed, and calculations have shown that the chances of a collision that would have this effect are practically zero.

All theories for the moon's existence have come up short. There are no processes observed today that can explain the existence of the moon without a Creator. The only idea that has not been shown inadequate is the one in the Bible. God created the moon and placed it in orbit around the earth. ■

WHAT DID WE LEARN?

- What are four secular theories for the origin of the moon?
- Which of these theories is most likely to be true?
- What does the Bible say about the origin of the moon?

TAKING IT FURTHER

- What are the main difficulties with the Capture Theory?
- Why do you think scientists come up with unworkable ideas for the moon's origin?

ORIGIN OF THE MOON

Read the following verses and discuss what each says about the origin of the moon.

- Genesis 1:14–19
- Psalm 8:3–4

- Psalm 33:6
- Psalm 74:16
- Psalm 136:3–9
- Jeremiah 31:35

It is clear from the Bible that no naturalistic explanation will adequately explain the moon's origin because its creation was supernatural.

UNIT 4

PLANETS

Pluto

Saturn

Jupiter

Mercury
Venus
Earth
Mars

Neptune

Uranus

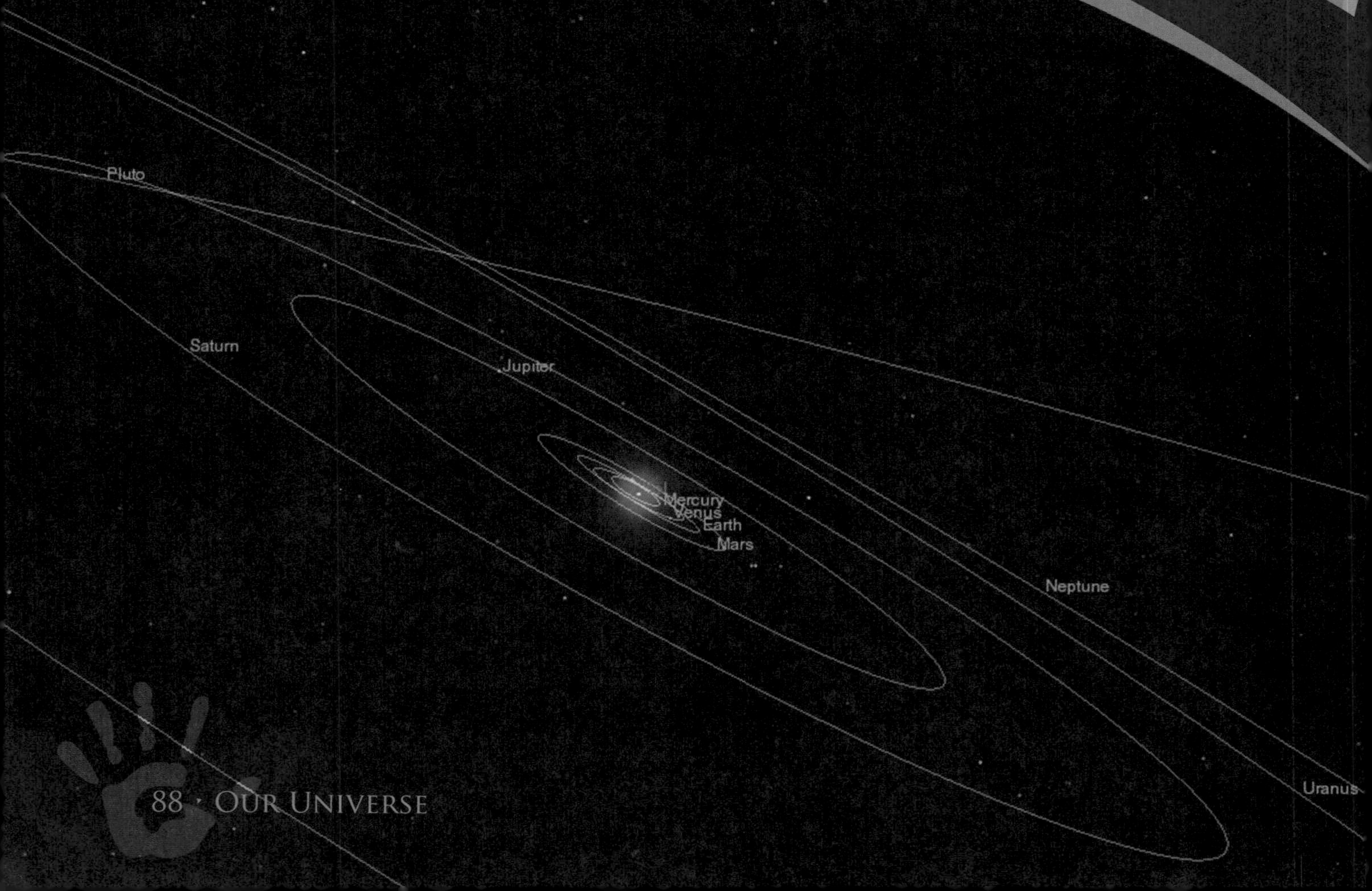

MERCURY

Closest planet to the sun

LESSON 19

What is the planet closest to the sun like?

Words to know:

terrestrial

Jovian

inferior planets

BEGINNERS

There are eight planets that move around the sun. Two of those planets are closer to the sun than the earth is, and the rest are farther away. The closest planet to the sun is called Mercury. Mercury is much smaller than the earth, and is the smallest planet in the solar system.

Mercury has very little atmosphere; that means it has no air around it. Because Mercury is so close to the sun, the side facing the sun gets very, very hot. And because it has no atmosphere, the side facing away from the sun gets very, very cold. You wouldn't want to visit Mercury.

• **Name three things you learned about Mercury.**

Our solar system has eight known planets that orbit the sun. All of the planets have elliptical orbits—orbits that are stretched circles. Those with solid surfaces are called terrestrial planets and those with gas surfaces are called Jovian planets, or gas giants. The planet with the closest orbit to the sun is Mercury. Mercury's average distance from the sun is 36 million miles (58 million km). Compared to the earth, Mercury revolves very quickly around the sun, making one complete revolution every 88 earth days. But compared to the earth, it rotates slowly on its axis, making one complete rotation every 58.6 earth days.

Mercury is a terrestrial planet. Its surface is covered with craters. This is most likely due to meteorites. Mercury is a small planet and its mass is much smaller than that of the earth; therefore, it has much less gravity and thus has almost no atmosphere to protect it from meteors like the earth does. What little atmosphere it does have is composed of helium and hydrogen. Its lack of atmosphere also allows for extreme temperature changes on the surface of the planet. The side of the planet facing the sun can be as hot as 800°F (470°C) and the side facing away from the sun can be as cold as −360°F (−180°C). Without a substantial atmosphere, Mercury does not have any weather. It was believed that Mercury was too hot to have any water, but in 1991, a small amount of ice was discovered in very deep craters near the poles where sunlight does not reach.

Mercury was named using Roman mythology. In Roman mythology, Mercury was the messenger of the gods. The planet Mercury is the smallest with a diameter of 3,031 miles (4,877 km). Its gravity is only 0.38 times as much as the gravity on earth. This means if you weigh 100 pounds on earth, you would weigh only 38 pounds on Mercury. Mercury does not have any moons.

Mercury and Venus are called interior planets because they are closer to the sun than the earth is. Sometimes they are also referred to as inferior planets. This does not mean there is something wrong with them, just that they have smaller orbits than the earth. From the earth, the interior planets appear to have phases just as the moon does.

For more information on Mercury and how it shows God's handiwork, see www.answersingenesis.org/go/mercury. ∎

FUN FACT

All of the craters on Mercury are named for artists, musicians, and writers. All of the valleys are named after observatories. And the ridges and cliffs are named after ships that have explored the earth.

WHAT DID WE LEARN?

- How do Mercury's revolution around the sun and rotation on its axis compare to that of the earth?

- What is the surface of Mercury like?

TAKING IT FURTHER

- How does a lack of atmosphere affect the conditions on Mercury?

ATMOSPHERE TEST

God provided a special atmosphere for the earth that allows life to flourish. Our atmosphere protects us from most meteors. It provides oxygen for us to breathe, circulates the air on the planet, and provides weather so we can grow food. The atmosphere also protects us from the extreme temperatures of space. Mercury's atmosphere, on the other hand, is too thin to protect it from extreme temperatures.

Purpose: To understand how the atmosphere protects the earth and how a lack of atmosphere affects Mercury

Materials: towel, hair dryer, ice

Procedure:

1. Wrap a towel around one arm.

2. Using a hair dryer on low heat, blow warm air on both of your arms. Compare the temperatures felt by each arm.

3. Next, place a piece of ice on each of your arms for a few seconds. Again, compare the temperatures felt by each arm.

Conclusion:

The towel insulates your arm from the hot air and the cold ice just as the earth's atmosphere insulates our planet from the heat of the sun and the cold of space. Without an atmosphere, Mercury's surface bakes in the sun and freezes in the shade.

MERCURY PROBE

For many years very little was known about Mercury. Scientists could see the planet with their telescopes, which gave them some idea of what the planet was like, but this did not answer all of their questions. Then in 1974, NASA was able to send a space probe called *Mariner 10* near Mercury where it was able to take photos and many scientific measurements of the planet. In one year, the *Mariner 10* was able to pass by the planet three times at three different levels, which allowed scientists to obtain valuable information about Mercury.

Scientists obtained some surprising results. First, they expected that any atmosphere Mercury may have had would have boiled away long ago, but instead, the probe detected a very thin atmosphere around the planet. Also, scientists were surprised to find that Mercury has a magnetic field. Because Mercury rotates slowly on its axis, scientists did not think that it could generate a magnetic field, yet it has a definite magnetic signature. Scientists are not sure how this field is generated.

The probe revealed that in many ways, Mercury is similar to our moon. Both objects are covered with craters. Also, both Mercury and the moon are believed to have small amounts of frozen liquid at the poles where the sun's rays do not reach. One distinct difference between the moon and Mercury however is their densities. Mercury

is much denser than the moon. Its density is similar to the density of earth.

Purpose: To help appreciate how little we can tell about a planet just by looking at it through a telescope

Materials: two index cards, crayons or markers, box or stack of books, tape, flashlight, magnifying glass, two clear plastic cups, water, milk

Procedure:

1. On an index card, draw a 1 inch circle and color it with many dark spots representing the craters on Mercury.

2. Set a box or a stack of books on a table and tape the index card to the box so that it is perpendicular to the table and its bottom edge is against the table.

3. Set a flashlight on the table so that its light is shining on the circle at an angle.

4. Hold a magnifying glass a few inches away from the index card on the opposite side from the flashlight.

5. Place another index card on the side of the magnifying glass opposite the original index card.

6. Move the card and the magnifying glass until you project a clear image of the planet onto the card.

This set up is similar to what a scientist does to view something in space. The flashlight represents the sun and the lens in the magnifying glass represents the lens in a telescope. The card on which the image is projected represents film from a camera or electronic sensor that detects the image and displays it. This set up is not complete however. Mercury has a very thin atmosphere. So let's make it a little more realistic.

7. Place a clear plastic cup between the flashlight and the index card with Mercury on it.

How did this affect the image on the second card? It probably blurred it slightly. The plastic causes some of the light rays to go in different directions so the image is not as bright or clear.

8. Now add a second plastic cup between the first index card and the magnifying glass to represent the atmosphere of the earth. How did this second cup affect the image? Again, it is probably slightly blurred and dimmer than the original image. Closely observe the image on the card. What can you tell about the planet just from looking at the image? Not as much as you would like. This is why space probes are so important for understanding the objects in our solar system.

As you will learn in our next lesson, Venus has a very thick opaque atmosphere.

9. Add water and two tablespoons of milk to the cup between the flashlight and the index card. How did this affect the image that is projected? It should block out the image.

Conclusion: Scientists cannot see the surface of Venus at all with their telescopes and have relied on space probes for pictures of its surface.

FUN FACT

In the mid-1800's a French astronomer named Urbain Leverrier believed that another planet had to exist. He believed that the orbit of Mercury was affected by a planet that had to exist between Mercury and the sun. He even named this planet Vulcan. However, no one has ever observed this "missing" planet.

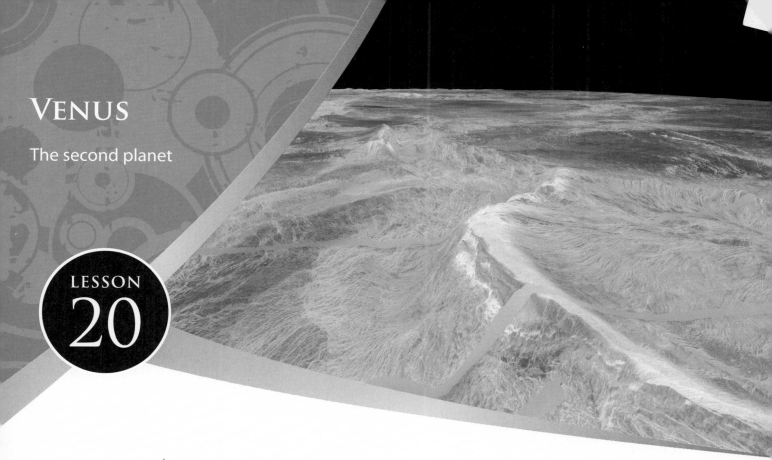

VENUS

The second planet

LESSON 20

What is the second planet from the sun like?

Words to know:

retrograde rotation

BEGINNERS

Venus is the second planet from the sun. Venus is only a little smaller than the earth, but you could not live on Venus. The planet's air is made of carbon dioxide, and it is surrounded by thick clouds of sulfuric acid. It would be poisonous for you to breathe the air. Also, because the atmosphere is so thick, it traps heat from the sun. Venus is the hottest planet in the solar system. You would burn up if you went there.

There is one good thing about the clouds that surround Venus. They reflect the light of the sun very well. This causes Venus to look like a very bright star in the night sky. Venus is often one of the first things you can see in the sky at night and one of the last things to disappear when the sun begins to rise, so it is sometimes called the *evening star* or the *morning star*.

- What is the name of the second planet from the sun?

- Why couldn't you live on Venus?

- Why can we see Venus in the night sky?

The second planet from the sun is Venus. This planet was named for the Roman goddess of love and beauty. Venus is sometimes called the *evening star* or the *morning star* because, besides the moon, it is often one of the brightest objects in the nighttime sky; thus it may be the first "star" to appear at sunset or the last "star" to disappear at sunrise.

Venus appears so bright in the sky because it has an atmosphere that reflects over 75% of the light that hits it. This atmosphere is 100 times thicker than earth's atmosphere and is made up of 90% carbon dioxide and about 10% nitrogen. In addition, the sky is full of clouds made of sulfuric acid. Although this atmosphere protects the planet from the extreme temperatures of space, because it is so thick it traps the sun's rays causing the surface to heat up. In fact, Venus is the hottest planet in the solar system, with a surface temperature of up to 900°F (470°C). This shows us that God provided just the right atmosphere for earth. Mercury has basically no atmosphere at all and Venus has an atmosphere that is poisonous to humans and keeps the planet too hot. In fact, none of the planets except earth has an atmosphere that would support life.

Venus's thick atmosphere exerts great pressure on the surface of the planet. Several space probes that were sent to investigate Venus were crushed by the atmospheric pressure as they approached the surface. Despite Venus's thick atmosphere, we have some idea of what its surface looks like. The *Magellan* spacecraft was the first space probe to be launched by the space shuttle. It was launched in 1989, and reached Venus four years later. It was not crushed and was able to take pictures of the surface of Venus. It revealed that the rocky surface has several large craters. Its atmosphere protects it from small meteors but large ones still occasionally make it to the surface. Also, we see evidence of former lava flows. This indicates that at one time there was extensive volcanic activity on Venus. It is believed that at one time the surface was covered with hundreds of volcanoes.

Because Venus is an interior planet, it reflects the light of the sun to earth. Depending on its relative position, we can see all, part, or none of the planet reflecting this light. So just like the moon and Mercury, Venus appears to have phases. In fact, this was one piece of evidence that helped convince people that the sun, not the earth, was the center of the solar system.

Venus is the planet that is closest in size to the earth. Its gravity is 0.91 times that of earth. Venus makes one revolution around the sun in 224.7 earth days and rotates on its axis once every 243 earth days, making Venus the planet with the slowest rotation in the solar system. Most planets rotate from west to east, but Venus rotates backwards, from east to west. This backwards rotation is called retrograde rotation.

THE GREENHOUSE EFFECT

The thick atmosphere on Venus acts like a greenhouse, trapping the sun's energy and heating the surface of the planet. To demonstrate this effect, follow the directions and complete the "Greenhouse Effect" worksheet.

Venus does not have any moons orbiting it. Its average distance from the sun is 67 million miles (108 million km).

For more on Venus and how it points to a Creator, visit www.answersingenesis.org/go/venus. ■

WHAT DID WE LEARN?

- Where is Venus's orbit with respect to the sun and the other planets?
- What makes Venus so bright in the sky?
- What is a nickname for Venus?
- How many moons does Venus have?

TAKING IT FURTHER

- Even though Venus has an atmosphere, why can't life exist there?
- Why doesn't the earth's atmosphere keep our planet too hot?

SURFACE MAPPING

As you learned in the last lesson, it was impossible for scientists to view the surface of Venus with telescopes. However, this did not mean that scientists had no knowledge of Venus's surface. Radio waves can penetrate the atmosphere around Venus and return information about the surface. In 1978, a probe called *Pioneer Venus 1* was sent to Venus. It used radio waves to map the surface of the planet. You can simulate this process by doing the following experiment.

Purpose: To understand how scientists can map the surface of Venus

Materials: shoebox, modeling clay, tape, graph paper, string, metal washer, ruler

Procedure:

1. In a shoebox, make a landscape with modeling clay. Include mountains, plains, and valleys. Be sure that the landscape does not go above the edge of the box.

2. Tape a piece of graph paper to the side of the box so that the top of the paper is even with the top of the box.

3. Make a probe by tying a metal washer on the end of a piece of string.

4. Use a marker and ruler to make marks every ½ inch on the string. Begin at the left hand edge of the graph paper.

5. Without looking into the box, lower the weight into the box until it hits something. Note which mark on the string is closest to the top of the box.

6. Remove the probe and measure how far the probe went into the box. Make a dot on the graph paper that distance from the top of the paper. For example, if the probe hit the surface of the planet 1½ inches down from the top of the box, the dot should be placed 1½ inches from the top of the paper.

7. Move to the next mark to the right on your graph paper and repeat this process. You should end up with a series of dots on the paper that when connected together resemble the landscape inside the box.

Conclusion:

From the information sent back by the radio probes, scientists discovered that Venus has mountains, valleys, plains, and craters similar to those on earth. They saw that Venus also has what are believed to be volcanoes. They even found that Venus has two large land masses similar in size to Australia and Africa. However, Venus does not have any water. Instead of vast oceans, Venus has dry, barren plains. As much as Venus may resemble earth in its size and form, it is very different in its ability to support life. God created a very special place for us.

Planets

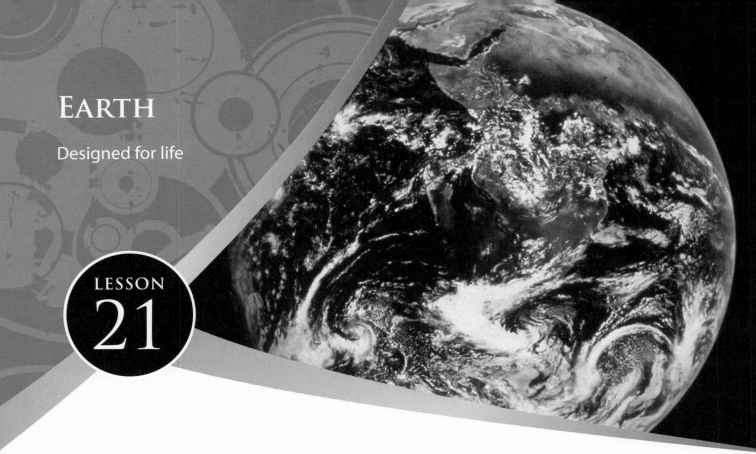

EARTH

Designed for life

LESSON 21

What makes our home, the third planet, special?

Words to know:

satellite

BEGINNERS

Earth is the planet that God designed just for people. God created everything just right so that plants, animals, and people could live here. It is just the right distance from the sun to keep us warm without cooking us. It has lots of water—which plants, animals, and people all need to live. No other planet has a significant amount of water on it. Earth has air with just the right amount of oxygen for us to breathe.

Earth is the third planet from the sun. It revolves around the sun once each year, and it rotates on its axis once each day. The earth has one moon that moves around the earth about once each month. The earth is a very exciting planet because God made it just for us.

- What is the third planet from the sun?

- List three ways that earth is just right for life.

- How long does it take for the earth to make one trip around the sun?

- How long does it take for the earth to rotate once on its axis?

Only one planet in the solar system can sustain life—earth. Our planet earth was designed by God to be the perfect place for us to live. However, as a result of Adam's sin in the Garden of Eden, God cursed the earth (Genesis 3); and later, because of the wickedness of all of the people of Noah's time, God judged the world with a global Flood (Genesis 6–9). This resulted in a planet that is much different from the paradise God initially created for man. However, the earth today is still a marvelous planet and a wonderful place to live!

Earth is the third planet from the sun. It revolves around the sun every 365.26 days, or one year, and rotates on its axis every 23 hours, 56 minutes and 4 seconds, or one day. It is tilted on its axis at 23 degrees from vertical with respect to the sun. The earth is an average of 93 million miles (150 million km) from the sun. All of this places earth in the perfect position for life.

Earth is the only planet in our solar system with a significant amount of water. About 72% of the surface of the earth is covered with water. This water is necessary for life to exist. Many scientists are searching for signs of water on other planets to see if life could have existed on other planets in the past. So far, there is no indication that life exists anywhere in our universe except on earth.

The earth has an atmosphere consisting of 78% nitrogen, 20% oxygen, 0.9% argon, and 0.1% carbon dioxide and other gases. As we learned in other lessons, this atmosphere protects the earth from the harsh temperatures of space and from many meteors that would otherwise hit the surface of the earth. But the atmosphere is special for other reasons, too. The closest layer to the surface of the earth, the first

FUN FACT

There is enough water in the oceans that if all the surface features of the earth were evened out, water would cover the earth to a depth of 1.7 miles (2.7 km).

EARTH MODEL

Purpose: To demonstrate why landmasses on flat maps often look different from the actual land-masses

Materials: globe, world map, orange, marker

Procedure:

1. Look at a globe of the earth. Notice the landmasses and the oceans. Consider how much water is on the earth and how it is needed to support life.

2. Now compare the globe to a world map. See how the landmasses and oceans near the poles have been stretched to change a round object into a flat map.

3. To make your own world map, draw the landmasses onto an orange with a marker.

4. After the marker has dried, care-fully peel the orange, in one piece if possible, and flatten it to make a map.

Conclusion:

Notice how the ends of the orange peel had to be broken apart to make the peel lie flat. This is why Greenland and other land masses near the poles look larger on many maps than they really are. If possible, locate a Merca-tor or cylindrical projection map of the earth and compare those with the map you have.

Now enjoy a delicious snack that could only be grown on planet earth.

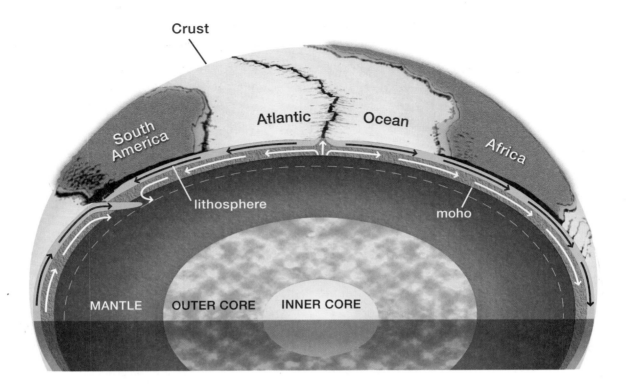

Crust

South America

Atlantic Ocean

Africa

lithosphere

moho

MANTLE   OUTER CORE   INNER CORE

0–10 miles (0–16 km), is called the troposphere. The troposphere is where the weather occurs. The weather patterns are vital to sustaining life on earth. The winds and air currents move water from the oceans where it evaporates, to the land where it can water plants. The weather also helps move cooler and warmer air around so the temperatures remain more constant around the earth. There is still a difference in temperature from one part of the earth to another. The temperatures can range from –60 to 140°F (–51 to 60°C), but this is a much smaller difference than the –360 to 800°F (–218 to 425°C) experienced on Mercury.

Earth is a terrestrial planet. It has a 5–35 mile (8–56 km) thick crust of rock. This crust floats on a mantle of melted rock that is about 1,800 miles (2,900 km) thick. The crust is broken into pieces called plates. These plates move slightly as they float on the mantel, and this movement sometimes causes earthquakes.

The earth has one natural satellite—the moon. The moon orbits the earth and reflects the light of the sun to give us light at night. The earth's moon is one of the larger moons in the solar system. God designed it just right to light up the night. Earth is God's special gift to humanity. ■

> ## FUN FACT
>
> At the equator, the earth is spinning at 1,000 mph (447 m/s) about its axis and moving at 67,000 mph (30,000 m/s) around the Sun.

WHAT DID WE LEARN?

- What are some features of our planet that make it uniquely able to support life?
- What name is given to the period of time it takes for the earth's revolution around the sun?
- What name is given to the length of the earth's rotation on its axis?
- On average, how far is the earth from the sun?

TAKING IT FURTHER

- What are some possible reasons why large amounts of water are found on earth but not on other planets?
- Why is it important that earth is a terrestrial planet?

 # WHY IS THE SKY BLUE?

Every child knows that on earth the sky is blue. This is true even from space. Pictures taken from space show that our planet is a beautiful blue color. So why is earth a blue planet?

Purpose: To demonstrate why the earth appears blue

Materials: two clear cups, milk, flashlight

Procedure:

1. Fill two clear cups with water.

2. Add a few drops of milk to one cup and stir the water just to mix the milk into it.

3. Take both cups to a dark room and shine a flashlight through each cup.

Questions:

- What color does the water appear to be in each cup? Why do you think this is?

Conclusion:

In the cup without milk, the water appears to be clear. In the cup with milk, the water has a pale blue color. This is because the milk droplets in the water disperse the blue light just as the air molecules disperse the blue light in our atmosphere. Our planet appears blue because the molecules in the atmosphere scatter the blue light waves more than they scatter other colors of light. This allows us to see blue more than other colors. Mars is called the red planet. Do you think

this is because of the scattering of red light in its atmosphere? You will find out in the next lesson.

MARS

The red planet

How is the fourth plant different from earth?

Words to know:

superior planets

BEGINNERS

Mars is the fourth planet from the sun. Mars is about half as big as the earth, making it just a little bigger than Mercury. Mars is sometimes called the red planet because the soil has a large amount of rust in it, giving the planet a reddish tint.

Mars has a small amount of air around the planet, but the air is mostly carbon dioxide, so you couldn't breathe it. Just like on earth, Mars has a north pole and a south pole. Both of these poles have ice on them all the time. But the ice on Mars is different from the ice on earth. Most of the ice on Mars is made from carbon dioxide, not water. Frozen carbon dioxide is called dry ice because when it warms up, it does not become a liquid but changes directly into a gas.

Most of what we know about Mars has been discovered by space probes that have visited the planet.

- What is the name of the fourth planet from the sun?

- How big is Mars compared to earth?

- Why is Mars sometimes called the red planet?

More than any other planet, Mars has evoked images of aliens and ideas for science fiction novels. Mars is the fourth planet from the sun and the first of the superior planets—those with larger orbits than the earth. Mars was named for the Roman god of war, perhaps because of its red color. It has two small moons called Phobos and Deimos, named for the attendants of Mars.

Mars is about half the size of the earth. It has a diameter of 4,222 miles (6,793 km) compared to earth's diameter of 7,927 miles (12,757 km). The gravity on Mars is nearly the same as the gravity on Mercury and is 0.38 times the gravity on earth.

Although Mars is small and has little gravity, it has a very thin atmosphere. Its atmosphere is about 7% as thick as the atmosphere around the earth and consists mostly of carbon

Planets

 ## EXPERIMENTING WITH POLAR ICE CAPS

NOTE: Dry ice must be handled by an adult wearing gloves! Never touch dry ice with your bare skin.

The polar ice caps on Mars are made mostly from frozen carbon dioxide, better know as dry ice. You can have some fun with dry ice and learn about Mars at the same time.

Purpose: To understand the properties of dry ice

Materials: dry ice, gloves, empty aquarium, candle, matches or lighter, cup of water

Procedure:

1. Using gloves, place a piece of dry ice in an empty aquarium or other similar tank. Observe the "smoke" coming off of the surface of the ice.

2. Light a candle and scoop some of the carbon dioxide gas from the tank and pour it over the candle. Observe what happens.

3. Carefully drop a small piece of the dry ice into a cup of water.

Observe the water "boil" as the ice turns to gas.

Questions:

• What was the "smoke" coming off of the dry ice?

• Why did the candle flame go out?

• Why did the water in the cup "boil"?

Conclusion:

Dry ice changes from solid to gas without going through a liquid phase. This process is called sublimation and is what caused the "smoke." In August 2003, Mars was closer to the earth than it has ever been. Observers using telescopes were able to see evidence of the polar ice cap sublimating. The gas brightly reflects sunlight causing the ice cap to appear brighter than the rest of the planet. Since dry ice goes directly from a solid to a gas without

leaving a liquid behind, it is used for transporting many frozen foods. It is also more efficient because it keeps foods colder than water ice does.

What happened to the flame? Fire requires oxygen. The gas from the dry ice is carbon dioxide. Carbon dioxide is heavier than air so when it is poured over the candle it pushes the oxygen molecules away, causing the fire to go out.

Carbon dioxide gas is what gives soda pop its fizz. The gas is added to the liquid under pressure. When a can of soda is opened, the gas begins to come out of the liquid, making it fizzy, just like your cup of water.

Although the polar ice caps on Mars may look like regular ice, the properties of carbon dioxide are very different from those of water and would not be able to support life.

dioxide. Mars has polar ice caps that grow in the winter and shrink in the summer. However, these ice caps are not made of water but are made from frozen carbon dioxide. Carbon dioxide freezes at a much lower temperature than water does and melts at temperatures above −110°F (−79°C). The thin atmosphere does not protect the planet from harsh temperatures, which range from −220 to 80°F (−140 to 27°C).

The surface of Mars is covered with many craters, mountains and valleys. Some valleys on Mars are much bigger than Grand Canyon on earth. The soil contains a high amount of iron oxide (rust), which gives Mars its characteristic red color. Periodically, the warming of the sun causes winds to blow, creating giant dust storms. Eventually the storms block out enough of the sun that the temperatures even out and the wind stops, allowing the dust to settle.

We know a great deal about Mars, due mainly to the Viking landers and other space probes, such as NASA's rovers, *Spirit* and *Opportunity*, that visited there and sent back information. The Mars rovers have found some evidence that liquid water may once have existed on Mars, but no probe has ever found any signs that life ever existed on the planet. Scientists continue to search for signs of life, and NASA plans to continue sending probes to learn as much as possible about the red planet. ∎

WHAT DID WE LEARN?

- Why is Mars called a superior planet?
- Why is Mars called the red planet?
- How many moons does Mars have?

TAKING IT FURTHER

- What causes the dust storms on Mars?
- Why doesn't the wind on earth cause giant dust storms like the wind on Mars?
- How would your weight on Mars compare to your weight on Mercury?

MARS PROBES

More space probes have visited Mars than any other object in space. Research some of the probes that have gone to Mars and what they have discovered. You may want to start by researching the Viking landers, but be sure to find out about more recent probes to the red planet, such as *Spirit* and *Opportunity*, as well. The NASA web site is a good place to start your research. Prepare a short presentation on what you learned about Mars space probes.

Planets

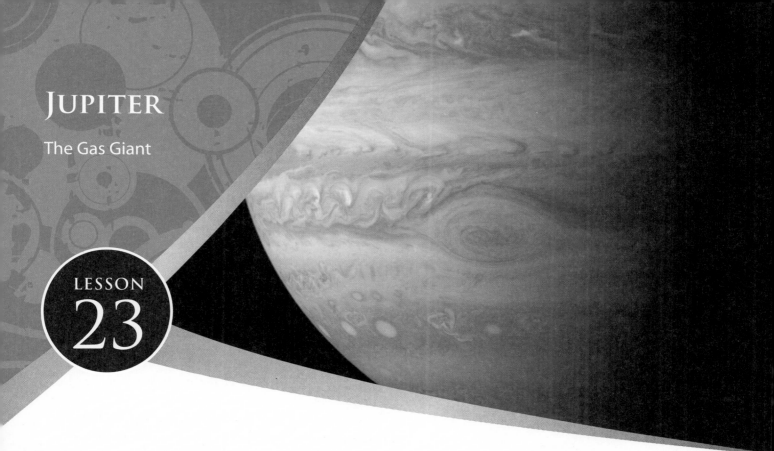

JUPITER

The Gas Giant

What is the largest planet in our solar system like?

Challenge words:

vortex

BEGINNERS

The fifth planet out from the sun is Jupiter. Jupiter is the biggest planet in our solar system. It is much, much bigger than earth. All of the planets that you have learned about so far are made of rock. But Jupiter and the other planets that are farther from the sun are different. They are giant balls of gas. There is no solid land on Jupiter.

One of the most famous features of Jupiter is its Great Red Spot. The Great Red Spot is believed to be a giant storm that has been raging for hundreds of years. People have known for many years that Jupiter has 4 large moons and at least 12 small moons that travel around it, but recent discoveries have shown that there are actually at least 60 moons that orbit Jupiter.

- **What is the name of the fifth planet?**
- **Name three things that are special about Jupiter.**

Mercury, Venus, Earth, and Mars are considered the terrestrial or "earth-like" planets. These planets are relatively small compared to the other planets in our solar system, and their surfaces are solid rock. Beginning with Jupiter, the outer planets are called Jovian or "Jupiter-like" because Jupiter, Saturn, Uranus, and Neptune are large gas planets.

Jupiter is the giant of our solar system. Its diameter is 89,372 miles (143,800 km) across compared with earth's diameter of only 7,927 miles (12,757 km). It is 1½ times as big as all of the other planets put together. Because of its large size, it was named Jupiter after the ruler of the Roman gods. Its large mass gives it a surface gravity that is 2.64 times that of earth.

Jupiter is the fastest spinning planet in the solar system. It rotates on its axis every 9 hours and 55 minutes. This fast rotation causes the planet to bulge noticeably at its equator. However, it is not as speedy around the sun. Its revolution around the sun takes 11.86 earth years to complete. It is an average of 483 million miles (778 million km) from the sun.

Jupiter is composed mainly of hydrogen and helium. It has an atmosphere of hydrogen gas that is hundreds of miles thick. The temperature at the top of the clouds is believed to be –250°F (–157°C). One of the most striking features of Jupiter's atmosphere is the Great Red Spot. This spot is an area in the atmosphere the size of three earths. Most scientists believe the Great Red Spot is a giant windstorm. It shrinks and grows and its color changes from pink to bright red. But it is in the same position and has the same oval shape that is has had since it was first discovered almost 300 years ago.

Jupiter's surface is made of hydrogen liquid and gas, and is believed to be 10,000 miles (16,000 km) deep. The center of the planet is very hot. This heat stirs up the liquid and gas hydrogen in a similar way to the convection experienced on the sun.

For years it was thought that Jupiter had 16 moons orbiting it, but recent

THE GIANT PLANET

Purpose: To appreciate just how big Jupiter is

Materials: two cereal bowls, marbles

Procedure:

1. Place two empty cereal bowls together to represent Jupiter.

2. Fill the two cereal bowls with marbles. The marbles represent the size of the earth.

3. Count how many marbles are in the bowls.

Questions:

- How many marbles fit in the two bowls?

- How many actual earths do you think could fit inside Jupiter? (The answer is more than 1,300!)

Conclusion:

What other items could be used to represent different objects in the solar system? Jupiter's diameter is about 11 times bigger than the earth's diameter, and the sun's diameter is about 10 times bigger than Jupiter's. What can you find that is about 10 times bigger around than the cereal bowl? Perhaps you have a giant beach ball or other very large ball that could represent the sun. What could you use to represent the dwarf planet Pluto? Pluto's diameter is about 5 times smaller than the earth's. Perhaps a BB could be used to represent Pluto. Take a guess at how many BBs would be needed to fill both cereal bowls. Now you have an idea of how large Jupiter really is.

discoveries have pushed that number to more than sixty. Most of these moons are very small, but four are quite large. The four largest moons are often referred to as the Galilean moons because Galileo was the first person to see them. Galileo named these moons Io, Europa, Ganymede, and Callisto. Io (shown here) has many active volcanoes. Europa is covered with ice. Ganymede is the largest moon in the solar system and is larger than the planet Mercury. Callisto is also larger than Mercury and is made of ice and rock. Astronomers discovered the smaller moons with better telescopes and with space probes. The *Voyager* space probe also discovered a ring around Jupiter that had not been seen by telescopes. It is now believed that Jupiter has four rings. These rings are made of bits of rock and dust and were most likely a result of particles that were broken off of Jupiter's moons when they were struck by meteorites. Instead of flying off into space, Jupiter's gravity pulled the particles into orbit. ■

Jupiter's moon Io

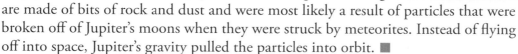

Planets

WHAT DID WE LEARN?

- What are some major differences between Jupiter and the inner planets?
- What is the Great Red Spot?

TAKING IT FURTHER

- Why does Jupiter bulge more in the middle than earth does?
- Why can't life exist on Jupiter?
- Why are space probes necessary for exploring other planets?

GREAT RED SPOT

Jupiter is well known for its Great Red Spot. Recently scientists have discovered a new spot on Jupiter that some scientists have dubbed "Red Spot Jr." Beginning in 2000, three smaller spots collided and formed one bigger spot. The original spots were white and for some time the new spot remained white. Then in 2003, the new spot began to change colors and is now almost identical in color to the Great Red Spot.

Scientists do not know why these storms are red. There are many ideas but no one knows for sure. Some scientists think that the power of the storm pulls material from deep below the clouds and lifts it up to the surface of the storm, where solar radiation somehow causes a chemical reaction that turns the storm red.

Purpose: To simulate the movement of particles inside the Great Red Spot

Materials: clear cup, tea bag, pencil

Procedure:

1. Fill a clear cup with water.
2. Open a tea bag and pour the tea leaves into the glass.
3. Insert a pencil into the center of the glass and swirl it around.

Conclusion: If you swirled the pencil properly, you created a **vortex**. Watch as the tea leaves are sucked into the center of the vortex. This is similar to how storms such as tornadoes and hurricanes work on earth as well.

SATURN

Surrounded by beautiful rings

LESSON 24

What makes the sixth planet unique?

Challenge words:

shepherding moons

BEGINNERS

Saturn is the sixth planet from the sun and the second largest planet in our solar system. It is also a gas planet like Jupiter. The most famous feature of Saturn is its rings. Since the invention of the telescope, scientists have been able to see something around the planet, but they did not always know what it was. Eventually, telescopes became good enough for scientists to see that there were rings around the planet. Space probes have shown that there are thousands of small rings around Saturn. These rings are made from bits of ice, dust, and rocks that orbit the planet.

Saturn also has at least 30 moons. One of these moons is large, six are medium-sized, and the rest are small. The largest moon is called *Titan*, and it is bigger than the planet Mercury.

- What are Saturn's rings made of?

- What is the name of Saturn's largest moon?

Saturn is the second largest planet in the solar system. Like Jupiter, it is a gas planet with an atmosphere composed mostly of hydrogen and helium. It is the sixth planet from the sun and was named for the Roman god of farming. However, Saturn is most famous for the beautiful rings that surround it.

Galileo was the first person to see the rings around Saturn when he viewed them with his telescope in 1610. However, he could not see them clearly enough to know what they were. Galileo thought that Saturn had two smaller globes circling around it because he saw what looked like "ears" on the sides of the planet. This phenomenon remained a mystery until 1659 when an astronomer named Christian Huygens was able to use a better telescope and discovered that the "ears" were actually rings around the planet.

From earth we can see only a few rings around Saturn. However, when the *Voyager* space probe explored Saturn in 1980, it was discovered that there are actually thousands of smaller rings around the planet. The rings are composed of pieces of ice, dust, and rocks. The band of rings is 170,000 miles (274,000 km) across and less than 3 miles (4.8 km) thick. Secular scientists have not been able to come up with a good explanation as to how Saturn's rings could form. Plus, we know that the rings are eroding quickly, so they could not have been there for millions of years. We know from the Bible that God created Saturn on Day 4 of Creation Week, about

Planets

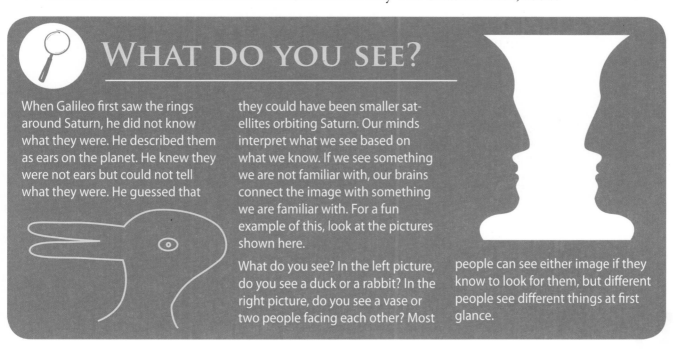

WHAT DO YOU SEE?

When Galileo first saw the rings around Saturn, he did not know what they were. He described them as ears on the planet. He knew they were not ears but could not tell what they were. He guessed that they could have been smaller satellites orbiting Saturn. Our minds interpret what we see based on what we know. If we see something we are not familiar with, our brains connect the image with something we are familiar with. For a fun example of this, look at the pictures shown here.

What do you see? In the left picture, do you see a duck or a rabbit? In the right picture, do you see a vase or two people facing each other? Most people can see either image if they know to look for them, but different people see different things at first glance.

> **FUN FACT**
>
> The Cassini-Huygens mission was launched on October 15, 1997. This is the first spacecraft to explore Saturn's rings and moons from orbit. The *Cassini* spacecraft entered orbit on June 30, 2004 and immediately began sending back intriguing images and data. The *Huygens* Probe dove into Titan's thick atmosphere in January, 2005 and sent data for about 90 minutes after reaching the surface. The touchdown on the surface of Titan marked the farthest a man-made spacecraft has successfully landed away from earth.

6,000 years ago, so the rings couldn't be older than that.

In addition to the thousands of rings, Saturn also has at least 30 moons, 12 of which were discovered in 2000, and according to recent NASA discoveries it is believed that Saturn may have as many as 56 moons. One moon is large, six are medium-sized, and the rest are very small. Most of the moons are covered in ice and full of craters. The largest moon is called *Titan*. Titan is the second largest moon in the solar system and is bigger than Mercury. Titan has an atmosphere consisting mostly of nitrogen with a small amount of methane. It is the only moon in the solar system known to have a substantial atmosphere.

Like Jupiter, Saturn spins very quickly on its axis. It makes one complete rotation every 10 hours and 40 minutes. This fast rotation causes Saturn to bulge in the middle just as Jupiter does. Saturn averages 887 million miles (1.4 billion km) from the sun and revolves around the sun once every 29.46 earth years. Saturn is tilted on its axis with respect to the earth in such a way that about every 15 years the rings are edge-on toward the earth, making them seem to disappear for a time. Saturn will line up in this way again in the year 2010. ■

WHAT DID WE LEARN?

- Who first saw Saturn's rings?
- What are Saturn's rings made of?
- What makes Titan unique among moons?

TAKING IT FURTHER

- Why did astronomers believe that Saturn had only a few rings before the *Voyager* space probe explored Saturn?
- Both Titan and earth have a mostly nitrogen atmosphere. What important differences exist between these two worlds that make earth able to support life but Titan unable to?

SATURN'S AMAZING FEATURES

Since the time of their discovery, Saturn and its rings have captured people's imaginations. There are seven major bands of rings around Saturn, labeled A through G. There is also a dark area between the A-ring and the B-ring that is called the Cassini Division. Each band of rings is from hundreds to hundreds of thousand of miles wide and can contain thousands of ringlets.

With the use of space probes, scientists have been able to learn more about these rings. In 1980, the *Voyager* space probe revealed that at least two of Saturn's moons, Prometheus and Pandora, may play a part in keeping the rings in place. These moons are very small. Pandora orbits on the outside of Saturn's F-ring while Prometheus orbits between the F-ring and Saturn. It is believed that the gravity from these moons plays some role in keeping the particles in the ring in orbit around Saturn. Thus, they have been called **shepherding moons**. However, there is some indication that these moons could also disrupt the orbit of some of the particles, so more study is needed to really understand what effect these moons actually have.

In 2004 the *Cassini-Huygens* space probe began sending back images from orbit around Saturn. These data have revealed many new and interesting things about Saturn. One moon, called Enceladus, which orbits in the E-ring, was observed shooting out jets of vapor and dust from its south pole. This is now believed to be the source of the material in the E-ring. Also, many new moons and moonlets have been discovered orbiting within the ring structures. There appears to be a somewhat complicated interplay between the moons and the rings. Data are still being collected and will continue to be analyzed in hopes of discovering

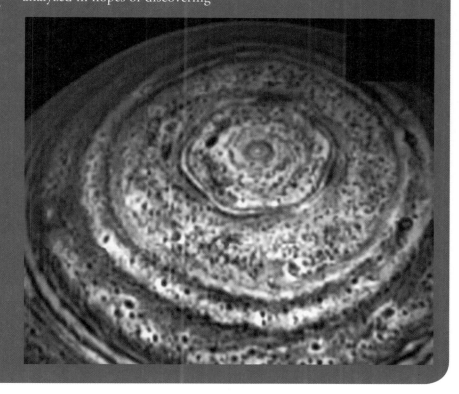

how all these things work together.

One amazing and unique feature discovered on Saturn is a double hexagon formed in the clouds of the planet's North pole (pictured below). Scientists are unable to explain its origin or how it has maintained this shape since it was first noticed more than 20 years ago. What do you think it might be?

While we may not understand all the features of Saturn, or the workings of its moons and rings, we can rejoice in the beauty that God has created.

Planets

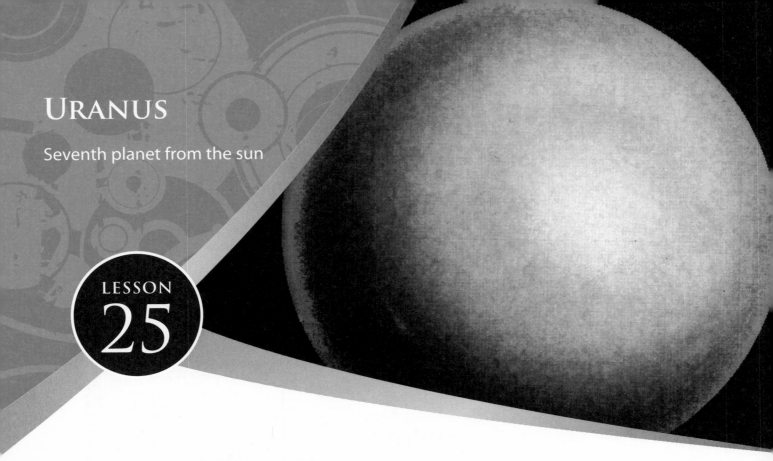

URANUS

Seventh planet from the sun

LESSON 25

What is the seventh planet from the sun like?

BEGINNERS

The seventh planet from the sun is Uranus (YOOR-uh-nus). Uranus is a gas planet like Jupiter and Saturn. It is smaller than Saturn but still much larger than earth. Uranus is a pale blue color.

The way that Uranus moves is different from other planets. Most planets spin like a top compared to the sun. If you put a stick through the center of most planets they would be mostly up and down compared to how the planets move around the sun. But Uranus rolls around the sun. A stick through its center would go from side to side.

You learned in the last lesson that Saturn has beautiful rings around it. Uranus also has rings; however, there are only a few rings around Uranus, while Saturn has thousands of rings. Uranus also has more than 20 moons, most of which are small.

- What is the name of the seventh planet from the sun?

- What color is Uranus?

- How does Uranus move that is different from other planets?

Even before the invention of the telescope, astronomers knew of the existence of five planets besides earth. The planets could be seen shining in the night sky. However, the planets farthest away were not discovered until more than 150 years after the invention of the telescope. The first of these new discoveries was made in 1781, when an English astronomer named William Herschel discovered Uranus (YOOR-uh-nus). He named the planet for the Roman god of the heavens.

Uranus is a pale blue color. It has an atmosphere of hydrogen, helium, and methane that completely obscures any view of the planet's surface. The atmosphere contains clouds and has winds that move from east to west at 90–360 mph (40–160 m/s). Its average distance from the sun is 1,783 million miles (2.8 billion km), and it revolves around the sun once every 84 earth years. Since it takes 84 years for Uranus to orbit the sun, each season on Uranus is 21 years long. This may sound like a long hot summer; however, the average temperature on Uranus is –365°F (–220°C), even in the sun. Uranus is so far from the sun, that the sun's rays have little effect on it so it remains very cold.

Uranus rotates on its axis once every 17 hours and 14 minutes. However, Uranus's rotation on its axis is unique. Whereas most planets rotate with a slight tilt from vertical with respect to the sun, Uranus rotates at a nearly 90-degree angle as if it were on its side compared to the other planets in the solar system. Uranus appears to roll around the sun because of this unusual tilt. This situation is impossible, according to evolutionary ideas about the formation of the solar system, namely that the planets condensed from a rotating nebula. Secular scientists have no plausible explanation for Uranus's strange angle of rotation.

Voyager 2 is the only space probe to visit Uranus. It visited the planet in 1986. Uranus has more than 20 known moons that orbit it. Ten of these moons were discovered by *Voyager 2*. Five of Uranus's moons are relatively large and the others

FUN FACT

Caroline Herschel, William's sister, became a noted astronomer in her own right—the first important woman astronomer. She discovered eight comets and three nebulae.

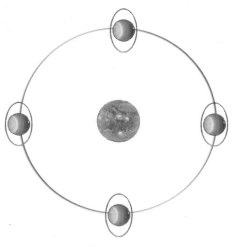

Planets

ROTATIONS & REVOLUTIONS

Purpose: To illustrate the unusual rotation of Uranus

Materials: ping-pong ball, two colors of paint, large ball

Procedure:

1. Paint one half of a ping-pong ball one color and the other half a different color.

2. Once the paint is dry, place a larger ball on the floor to represent the sun.

3. Practice rolling the ping-pong ball on the floor around the larger ball in such a way that it rolls along the line separating the two colors, so that one side of the ball is facing the "sun" at one side of the orbit and the other side is facing the "sun" when it is half way around the orbit, like the planet in the diagram above. This is how Uranus orbits the sun.

are relatively small. Several of the moons orbit very close to the planet. Just as the planet rotates at a 90-degree angle, so the moons revolve around the planet at a 90-degree angle from how the moons revolve around the other planets. The moons revolve around the planet vertically. The diagram on the previous page illustrates how Uranus orbits the sun and how the moons orbit Uranus.

In addition to the moons, Uranus is also surrounded by at least 13 rings. Some of these rings have been discovered using telescopes, and others were discovered when *Voyager 2* passed by Uranus. Some of the rings contain chunks of black material that has yet to be identified. Much remains unknown about this planet because it is so far away from the earth.

For more information on Uranus and how it shows God's handiwork, see www.answersingenesis.org/go/uranus. ■

WHAT DID WE LEARN?

- What makes Uranus unusual compared to the other planets?
- How have rings been discovered around Uranus?

TAKING IT FURTHER

- How can we learn more about Uranus?
- Why is Uranus such a cold planet?

TILT AND ROTATION

Each planet in the solar system orbits at a different angle compared to the plane in which it orbits the sun, and some planets rotate in opposite directions. Below is a chart showing the angle and direction of rotation for each planet. Use this information to make models of each planet.

Purpose: To make a model of each planet showing angle of tilt and direction of rotation

Materials: modeling clay, 8 pencils, protractor, index cards

Procedure:

1. Make a ball of clay to represent each planet.

2. Push a pencil through each ball to represent the axis of rotation.

3. Place a piece of modeling clay on the table and insert the sharpened end of the pencil into the clay.

4. Use a protractor to measure the angle of the tilt for that planet and press the clay against the pencil to hold it in place.

5. Write the name of the planet, its angle of tilt, and its direction of rotation on an index card and place the card in front of the model.

6. You can take a picture of your models to help you remember what they look like.

Planet	Angle of Tilt from Vertical	Direction of Rotation (viewed from North pole)
Mercury	0°	Counter clockwise (toward east)
Venus	3°	Clockwise (toward west)
Earth	23°	Counter clockwise (toward east)
Mars	24°	Counter clockwise (toward east)
Jupiter	3°	Counter clockwise (toward east)
Saturn	27°	Counter clockwise (toward east)
Uranus	98°	Clockwise (toward west, forward)
Neptune	29°	Counter clockwise (toward east)

NEPTUNE

Last of the gas giants

LESSON 26

What is the farthest planet from the sun like?

Challenge words:

centripetal force

BEGINNERS

Neptune is the eighth planet from the sun. It is about the same size as Uranus. Neptune is made of gas just like Jupiter, Saturn, and Uranus. Its atmosphere makes it look blue.

Neptune has at least four rings and thirteen moons. Two of the moons are large and the rest are small. The largest moon is called *Triton*. Triton is about the same size as our moon. Because Neptune is so far away from earth, we don't really know very much about it. In fact, Neptune is so far away from the sun that the sun would appear to be only a bright star from the surface of Neptune.

- What is the name of the eighth planet from the sun?

- What color is Neptune?

- Is Neptune a solid or a gas planet?

The last of the gas giants is Neptune, the eighth planet from the sun. Neptune was discovered in 1846, and named for the Roman god of the sea. It was discovered in a rather unusual way. Astronomers noticed that when they observed Uranus's location it was not always quite where they expected it to be. They theorized that the gravitational pull of another planet could be affecting its orbit. A British astronomer named John Couch Adams and a French astronomer named Jean Leverrier calculated where a planet would have to be in order to affect Uranus' orbit. When the German astronomer Johann Gottfried Galle looked in the predicted location, he discovered Neptune only one degree off from the predicted location.

Neptune is so far away that it can only be viewed with a very good telescope. It is an average of 2,794 million miles (4.5 billion km) from the sun. It is so distant from the sun that from the surface of Neptune, the sun would appear to be only a bright star. It revolves around the sun once every 164.8 earth years and rotates on its axis once every 18 hours and 30 minutes.

Neptune has an atmosphere composed mostly of hydrogen, helium, and methane. Methane absorbs red and other colors and reflects blue so the planet appears to be blue. It has streaky methane ice clouds. And like Jupiter, Neptune had what is believed to be a giant storm in an area called the Great Dark Spot. It is believed that the winds in the Great Dark Spot blew at more than 700 mph (313 m/s). The Great Dark Spot seems to have disappeared for now. But astronomers believe they discovered a new storm in 1994, using the Hubble Telescope. Some scientists believe that the Dark Spot may actually be

REFLECTION OF COLORED LIGHT

Purpose: To demonstrate how different colors of light can be absorbed

Materials: three clear cups, blue food coloring, red food coloring, flashlight

Procedure:

1. Fill three clear cups with water.

2. Add blue food coloring to one cup and red food coloring to another.

3. Shine a flashlight through the clear cup of water, projecting the beam onto a wall. What color is the light? (It should be white, or perhaps a bit yellow or blue depending on the type of bulb in the flashlight.)

4. Next, shine the flashlight through the cup with blue water. What color was the light on the wall? (The light was blue because the blue water absorbed all colors except blue and allowed the blue light to pass through.)

5. Now shine the flashlight through the cup of red water. What color was the light on the wall? (The red water absorbed all colors of light except the red light.)

Conclusion: Like Uranus, Neptune's atmosphere is primarily made up of hydrogen, helium, and methane. Methane absorbs all colors of light except blue so only blue light is reflected, giving the planet a blue appearance.

Similar to our experiment, scientists can determine the elements in a planet's atmosphere by the wavelength of light that is reflected.

a hole in the clouds surrounding Neptune. Until more information can be gathered we may not know for sure what these dark spots are.

Neptune is the farthest out of the gas giants. It is nearly the same size as Uranus. It has at least four rings and 13 moons. Two of the moons are large and the rest are small. The largest moon is called *Triton*. Triton is about 1,700 miles (2,736 km) across, nearly the same size as our moon. Triton has a retrograde, or backwards, orbit around the planet compared to the orbits of other moons around their planets. Triton has many geysers that shoot out cold nitrogen gas and dust particles.

For more information on Neptune and how it shows God's handiwork, see www.answersingenesis.org/go/neptune. ■

WHAT DID WE LEARN?

- What similarities are there between Uranus and Neptune?
- What are two possible explanations for the Great Dark Spot?

TAKING IT FURTHER

- Explain how Neptune was discovered.
- What affects the color of a planet?

CENTRIPETAL FORCE

Take a close look at the Planet Statistics on page 119. What do you notice about the length of time it takes for each planet to orbit the sun? You will notice that Mercury, the closest planet, obits the sun in only 88 days. Pluto, farthest from the sun, orbits the sun in 248 years, which is about 90,520 days. If you convert all of the orbits into days, you will see that the farther a planet is from the sun, the slower it orbits the sun. Let's see why this is.

Purpose: To see how gravity affects rate of orbit

Materials: metal washer, piece of string

Procedure:

1. Tie a washer to the end of a string that is three feet (1 m) long.

2. Go outside away from any buildings and other people to try this experiment. Holding the string near the end, twirl the string above your head. Exert just enough force to keep the washer moving at its slowest speed. The string is applying centripetal force to the washer much like the sun's gravity pulls on the planets and keeps them in orbit.

3. Next, hold the string in the middle and repeat the experiment. Note the pressure between your hand and the string and the speed you must spin the washer to keep it moving.

4. Now, shorten the string to only a few inches and spin the washer again. It should spin very quickly. Again, note the pressure between your hand and the string and the speed of the washer.

Questions:

- Were you able to spin the washer as slowly after you shortened the string?
- How does the pressure between your hand and the string compare when the string is short and when the string is long?

Conclusion: As the string shortened, you had to exert more pressure to keep the washer spinning and its speed increased. In much the same way, the gravitational pull is greater on a planet that is closer to the sun. The gravity of the sun exerts a force on each planet called **centripetal force**, which is the force that causes something to move in a circle. In our experiment, the pull of the string was the centripetal force. The greater the force the faster something moves. So as the gravity decreases it causes the planet to orbit at a slower rate.

PLUTO & ERIS

Dwarf planets

LESSON 27

What is Pluto, the former planet, like?

Words to know:

synchronous orbit

BEGINNERS

Pluto, once considered the ninth planet in our solar system, was reclassified as a dwarf planet in 2006. Another object that orbits the sun and is small like Pluto is also considered a dwarf planet. It is called *Eris*. Pluto is so far from the sun that the sun looks just like a bright star. Pluto is very cold all the time.

Pluto has a large moon called *Charon*. Charon is about half as big as Pluto. Charon orbits very close to Pluto. Recently, scientists also discovered two smaller moons orbiting Pluto. These moons are much smaller than Charon and orbit much farther away. These moons have been given the names *Nix* and *Hydra*.

- Is Pluto considered a true planet?

- What is the temperature like on Pluto?

- How many known moons does Pluto have?

- What is the name of the largest moon orbiting Pluto?

Pluto is very small and very far away from the sun. Because of its distance from earth and its small size, it was not discovered until 1930. A young American astronomer named Clyde Tombaugh is credited with the discovery. Although Percival Lowell, owner of the Lowell Observatory, believed that there was a ninth planet, he was unable to find it before he died. Tombaugh, an astronomer working at the Lowell Observatory in Arizona, used Lowell's calculations and eventually found the elusive planet. It was much dimmer than Lowell expected and was thus harder to find. The name *Pluto* was first suggested by Venetia Burney, an 11-year-old school girl from England. The name won out over numerous other suggestions partly because it was named after the Roman god of the underworld (who was able to make himself invisible) and because the first two letters were the same as Percival Lowell's initials.

Because Pluto is so far away, very little is known about it. It is believed to be a ball of frozen methane, water, and rock. It may possibly have a very thin methane atmosphere, but only part of the time. The gases become frozen when Pluto is far from the sun and only comprise an atmosphere when the planet is close enough to the sun to melt the gases. It is believed that the average temperature on Pluto is a cold −370°F (−223°C).

In 1978 a moon was discovered orbiting Pluto. It was named *Charon* after the ferryman of the dead in Greek mythology. Charon is about 50% the size of Pluto, making it very large in comparison to the object it orbits.

Some people have said that Pluto and Charon should be considered a double planet system. In fact, there has been considerable debate on how to define a planet and whether Pluto should even be considered one. In 2006 the International Astronomical Union, a large group of astronomers, decided to define a planet in such a way that Pluto is no longer classified as a regular planet. It was called a *dwarf planet*. This controversy may not be over, since many people feel strongly that Pluto should be considered a planet. What do you think?

Charon orbits Pluto in a synchronous orbit. This means that Pluto and its satellite are always presenting the same side to each other. If you were to stand on Pluto, Charon would appear to hover in the sky and not move. In May 2005 the Hubble telescope was being used to observe Pluto, and scientists discovered two new satellites orbiting Pluto. Additional images from the Hubble in 2006 confirmed that Pluto has two additional moons. These new moons have been named *Nix* and *Hydra*, also names from Greek mythology.

Pluto's orbit around the sun is an elongated ellipse that actually crosses Neptune's orbit. So, for 20 out of every 250 years, Pluto is closer to the sun than Neptune. The

HOW MUCH DO I WEIGH?

Complete the "How Much Do I Weigh?" worksheet.

last time this occurred was from 1979–1999. From Pluto, the sun appears to be a small bright dot in the sky.

Pluto is very small. Its diameter is only 1,425 miles (2,294 km), compared to earth's diameter of 7,927 miles (12,757 km). It has a very small mass and is believed to have a gravitational pull of only 0.08 times that of earth.

Another object that orbits the sun was discovered in early 2005. Eris, as it has been called, is estimated to be approximately 1,500 miles (2,400 km) in diameter, which is larger than Pluto, and its distance from the sun is nearly three times that of Pluto. It is also classified as a dwarf planet. Several planets have recently been discovered that orbit other stars, but they are so far away that we don't know much about them. They are called *exoplanets*. Beyond our solar system there could be millions of planets orbiting other stars. We do not yet possess the technology to see in detail beyond our solar system; however, we will continue to study the vastness of space and learn more about the wonderful universe God created. ∎

Planets

WHAT DID WE LEARN?

- What discovery was originally considered to be the ninth planet?
- How does the gravity on Pluto compare to the gravity on earth?
- Is Pluto always farther from the sun than Neptune?
- What is unique about how Charon orbits Pluto?

TAKING IT FURTHER

- Why did it take so long to discover Pluto?
- Why is Pluto no longer considered to be a planet?
- What alternate classification was given to Pluto in 2006?

NEW HORIZONS

Because the only way to get detailed information about a planet is to send a space probe there, NASA launched a probe called *New Horizons* on January 19, 2006. It is scheduled to pass Uranus in 2011, Neptune in 2014, and reach Pluto in 2015. *New Horizons* carries seven different scientific instruments that are designed to give us information about the composition, geography, and atmosphere of both Pluto and Charon. *New Horizons* will give us the best opportunity to observe the newly discovered moons as well.

Once the probe sends back information on Pluto and Charon, it is scheduled to continue on into the Kuiper Belt beyond Pluto's orbit and send back information on what it encounters there.

To find out where *New Horizons* is now, you can visit the NASA web site at http://pluto.jhuapl.edu/ mission/whereis_nh.php. This site shows a picture of where the probe is on its journey to Pluto and beyond.

Planet Statistics

One way to learn more about the planets in our solar system is to compare certain statistics. By seeing how the various planets compare to earth, we can get a better feel for how big the other planets are, where they are located compared to the earth, and how they move.

Planet	Avg. distance from sun (in millions)	Revolution around sun	Rotation on axis	Diameter	Volume compared to Earth	Mass compared to Earth
Mercury	36 miles (58 km)	88 days	58.6 days	3,031 miles (4877 km)	0.06	0.056
Venus	67 miles (108 km)	224.7 days	243 days	7,521 miles (12,101 km)	0.97	0.82
Earth	93 miles (150 km)	365.26 days (1 year)	23 hr 56 min (1 day)	7,927 miles (12,757 km)	1.0	1.0
Mars	143 miles (230 km)	628 days	24 hr 38 min	4,222 miles (6793 km)	0.15	0.11
Jupiter	483 miles (778 km)	11.86 years	9 hr 55 min	89,372 miles (143,800 km)	1324	318
Saturn	887 miles (1,427 km)	29.46 years	10 hr 40 min	74,990 miles (120,660 km)	736	95.1
Uranus	1,783 miles (2,870 km)	84 years	17 hr 14 min	31,700 miles (51,118 km)	64	14.5
Neptune	2,794 miles (4,497 km)	164.8 years	16 hr 6 min	30,764 miles (49,500 km)	58	17.2
Pluto (dwarf planet)	3,666 miles (5,900 km)	248 years	6 days 9 hr	1,425 miles (2,294 km)	0.01	0.002
Eris (dwarf planet)	9,000 miles (15,000 km)	557 years	8 hr	1,500 miles (2,400 km)	0.01	0.003

UNIT 5

SPACE PROGRAM

NASA

The National Aeronautics and Space Administration

What is NASA and what does it do?

Words to know:

NASA

escape velocity

Challenge words:

NACA

supersonic

BEGINNERS

If you are at all interested in space travel, you have probably heard of NASA. The letters in NASA stand for National Aeronautics and Space Administration. NASA is a group of people who work on many different things that have to do with space.

The people at NASA design spacecraft like rockets and the Space Shuttle. They also design space probes that visit other planets. NASA trains astronauts to work in space, and they are the people who are building the space station. Some people at NASA plan the flights of the rockets and shuttles. Others helped build the Hubble Space Telescope.

The first major job that NASA had was to find a way to send a man to the moon. NASA did this in 1969. Since then, many men have been to the moon and even more people have been to outer space in a space shuttle or in the space station.

Another important thing that people at NASA do is to study the information that is sent back by probes and to conduct experiments in space. Many astronauts do important experiments on the space station. Why do you think they do the experiments there instead of on earth? It is because there is virtually no gravity on the space station, so things act differently on the space station than on earth.

- **What is NASA?**

- **Name three ways that NASA studies things in space.**

- **What was the first important job that NASA had to do?**

The study of space is a very exciting field, and although many countries and many scientific groups have contributed to our exploration of space, none has done as much to promote our understanding of the universe as NASA has. NASA, the National Aeronautics and Space Administration, is a government-funded scientific organization dedicated to exploring the universe.

NASA was founded in 1958 by President Dwight Eisenhower. It was an outgrowth of the National Advisory Committee on Aeronautics, an organization that was already working to improve aircraft. NASA's focus was not only on flying vehicles, but also on space exploration.

President Dwight Eisenhower

Throughout the 1960s NASA's main goal was to develop the technology necessary to place a man on the moon. This was accomplished through three separate programs: Mercury, Gemini, and Apollo. We will study these programs in more detail in a later lesson. In addition to these programs, NASA also worked on weather and communications satellites.

Today, NASA is divided into four main groups. The aeronautics research group develops better design tools and technologies for improving vehicle and air system safety and performance. The ideas from this group are used not only to improve NASA vehicles, but are instrumental in improving commercial and military aeronautics as well. The second group is the exploration systems group. This group is dedicated to developing the vehicles and systems necessary to allow humans and robots to explore beyond the reaches of earth. The third group in NASA is the science group. The science group plans missions, conducts experiments, analyzes data, and much more. Finally, the fourth group is the space operations group. This group oversees all of what goes on in space. Its primary responsibilities include overseeing the International Space Station and the Space Shuttle flights.

As you can see, NASA is involved in a broad scope of activities, from research and development, to space flights and data analysis. Many of these activities have had a profound impact on everyone on earth. Not only are we more aware of what is in the universe, but many of the technologies developed for the space programs have contributed to non-space developments as well. In the 1960s NASA developed digital imaging systems to receive images from the moon. This technology led to medical imaging systems that are used to help diagnose and treat patients. More recently, NASA developed a cable system that responds to human touch. This system has been developed into a special kind of walker that aids patients with spinal cord injuries in their rehabilitation process. Another NASA invention is likely to benefit people at the gas pump. NASA developed a new flow meter process for the Space Shuttle. This meter is being made available to oil and gas refineries and other industries that must regulate the flow of liquids. The NASA flow meter is much more efficient than older systems and can greatly reduce the energy needed to push the liquids through the pipes. Other examples include the joystick control system used to maneuver the

ESCAPE VELOCITY

In order to launch a vehicle into space, scientists had to find a way for it to escape the pull of earth's gravity. Since the earth is always pulling things toward the center of the earth, a spacecraft must be moving fast enough to overcome this force. The speed required to overcome gravity is called escape velocity. On earth, escape velocity is approximately 25,000 mph (40,000 km/h). So far, all spacecraft that have been launched have used rockets to generate escape velocity. However, NASA is working with the aviation industry to develop jet technology that can also propel aircraft into space.

Purpose: To demonstrate how escape velocity works

Materials: tag board, tape, one magnet, plastic lid or dish, book, steel BB

Procedure:

1. Cut a 4-inch wide by 12-inch long strip of tag board or stiff paper. Fold the paper in half the long way. Now, fold each long edge of the paper back toward the fold so that the paper looks like an M. This is your ramp.

2. Tape a magnet to one end of the ramp.

3. Place the magnet end of the ramp inside the edge of a plastic lid or dish with sides.

4. Place a book under the other end of the ramp to give it a slight tilt.

5. Now, place a steel BB at the top of the ramp and let it go. Note what happens to it. If the BB sticks to the magnet, remove it.

6. Next, add another book under the end of the ramp to make it steeper. Release another BB from the top of the ramp. Note the speed of this BB. If the second BB sticks to the magnet, remove it from the magnet.

7. Add a third book to the ramp, and try it again. What happened to this BB?

Questions:

- What happened to the first BB? Why?

- What happened to the second BB? the third? Why?

Conclusion:

When the BB is moving faster it has more momentum; thus it takes more force to stop it. The first BB was probably attracted to the magnet because it was not moving fast enough to escape the magnetic pull. This is similar to the pull of gravity on a space craft. If the second or third BB was moving fast enough, the force of the magnet would not have stopped it. Similarly, if a spacecraft is moving fast enough, the earth's gravity does not have enough force to stop it and pull it back to the surface of the earth. This required speed is called the escape velocity.

lunar vehicle, which has been adapted to automobiles, allowing disabled people to operate their cars using only their hands, and the development of scratch-resistant eyeglasses. The space industry developed a strong yet light coating to protect equipment in space. This coating was later adapted as a coating for plastic lenses that provides greatly-improved scratch resistance. As NASA develops new technologies for space research, many of these technologies will be adapted to uses closer to home.

Although NASA has accomplished many great feats, it should be noted that NASA as a whole is dedicated to evolutionary ideas. Many of its programs are being designed specifically to prove that the origins of the universe and life are naturalistic. Because of their blatant evolutionary bias, NASA scientists will interpret data to fit their old universe, naturalistic worldview and will likely miss many of the wonders that God has waiting for us to discover. ■

WHAT DID WE LEARN?

- What is NASA?
- When was NASA formed?
- What was one of NASA's first tasks?
- List at least three of NASA's work groups.

TAKING IT FURTHER

- How does NASA help people who are not interested in space exploration?
- How might an evolutionary worldview affect NASA's work?

NACA

NASA grew out of an organization called the National Advisory Committee on Aeronautics (NACA). NACA was created by President Woodrow Wilson on March 3, 1915. According to President Wilson, its purpose was "to supervise and direct the scientific study of the problems of flight, with a view to their practical solution." NACA provided oversight and direction for the development of new aircraft from 1915 until it became part of NASA in 1958.

NACA was instrumental in developing many of the technologies that were used in civilian and military aircraft. NACA engineers designed better engines, better airfoils, and better wing designs. They also developed

Pilot Neil Armstrong next to the X-15 after a research flight

the first supersonic (faster than the speed of sound) wind tunnel to allow testing of designs for supersonic aircraft.

NACA also pioneered many of the inventions and ideas that have made flying safer. The NACA engineers designed a system to prevent ice formation on wings and propellers, and were the first to use a refrigerated wind tunnel to test their inventions under cold conditions. NACA also instituted licensing of pilots, supported weather research to improve navigation safety, and recommended inspection and expansion of airmail.

In 1952 NACA began doing research on problems that might be encountered in space. In 1954 it worked together with the U.S. Air Force to develop a high altitude research vehicle called the X-15 (shown here). This aircraft was a rocket-propelled craft that could fly into the upper atmosphere and give researchers the chance to see what space flight might be like.

In 1958 NACA became part of the new space agency, NASA, but they continued to do aeronautic research. Today, the scientists at NASA are working on technology that will allow aircraft to go into space using propulsion methods other than rockets. In 2004, NASA engineers set air speed records for air-powered engines by flying their plane at Mach 9.6 (6,800 mph).

NACA has always believed that sharing its technology with the civilian and military aircraft designers benefits everyone. Thus, the vision of President Wilson to find practical solutions to the problems of flight continues to live on in the work that is being done by the scientists and engineers at NASA today.

Questions:

- What was NACA?
- What was its original purpose?
- What were some of the major contributions to aeronautics that were made by NACA?

Space Program

SPACE EXPLORATION

Seeing what's out there

LESSON
29

How do we study
deep space?

Words to know:

space probe

Challenge words:

suborbital

BEGINNERS

Many people have dreamed of going into space, but until about 60 years ago this idea was little more than a dream. During World War II, rockets were first developed. Later, these rockets were improved until they were good enough to launch things into space. The first thing to be placed in space was a satellite called *Sputnik*. A man-made satellite is something that moves around the earth in space.

After satellites were successfully launched into space, special space vehicles and rockets were developed to launch people into space. The first man in space was a Soviet named Yuri Gagarin. Just a few months after Gagarin made his journey into space, the first American, Alan Shepherd, was launched into space. Both of these men went into space in 1961. After this, America began working on rockets and space ships that could take people to the moon. In 1969, Neil Armstrong and Buzz Aldrin became the first people to ever walk on the moon.

NASA has sent many different kinds of satellites into space. Many of these satellites take pictures of the earth or monitor the weather. Other satellites are used to send television and telephone signals from one place to another. One of the most important satellites is the Hubble Space Telescope, which can view and take pictures of the planets in our solar system as well as stars that are very far away.

NASA also builds and launches **space probes**. A probe is different from a satellite. A satellite stays in space around the earth, but a probe is sent to other planets. Most of what we know about the planets in our solar system has been learned from the information that space probes have sent back to us.

- **What is a satellite?**

- **What is a space probe?**

- **What invention was needed in order to send satellites and other items into space?**

Robert Goddard with an early rocket

Sputnik

The idea of space travel originated in science fiction. The earliest ideas were in the writings of Jules Verne, H. G. Wells, and other 19th-century authors. At that time no one took the idea of space travel seriously. However, Robert H. Goddard was influenced by these writers and began experimenting with rockets. He wrote his own book on rocket flight, called *A Method of Reaching Extreme Altitudes*, in 1919. Goddard performed the first flight of a liquid-fueled rocket on March 16, 1926. He continued his work on rockets and flew the first rocket to go faster than the speed of sound in 1935. Today, Robert Goddard is considered the father of modern rocketry. NASA's Goddard Space Flight Center near Washington, DC is named for him. However, during his lifetime people did not take Goddard's work seriously. It wasn't until after rockets were used during World War II that the idea of sending rockets into space took hold. After World War II, Werner VonBraun and other German scientists came to America to help develop rockets, and Soviet scientists began working on rockets in earnest as well. The earliest rockets were unreliable, but as rockets became more reliable, a race began to see who would be the first nation to get to space.

The Soviet Union (a former northern Eurasian empire from 1922–1991, consisting of 15 Socialist Republics of which Russia was the largest) won the first lap of the race by putting the first man-made satellite into orbit. *Sputnik*, shown here, was launched October 4, 1957. The first US satellite, *Explorer 1*, was launched just a few months later on January 31, 1958. This sparked a huge interest in space and rockets around the world.

The next step was to put a man into space. Again, the Soviets were the first to do this. Yuri Gagarin was the first man in space on April 12, 1961. The first American in space was Alan Shepherd in May, 1961, and the first American to orbit the earth was John Glenn on February 20, 1962. The Soviets continued to lead in the space race by putting the first woman in space when Valentina Tereshkova went up in 1963, and by performing the first space walk in 1965.

These early successes by the Soviet Union led U.S. President John F. Kennedy, in May 1961, to challenge America to place a man on the moon by the end of the decade. This was a monumental task to complete in less than nine years. This was accomplished through three projects. The Mercury project had the goal of manned space flight. Each Mercury capsule held one person. There were five orbital flights in all. The second project was called Gemini and was designed for two-man space flight. Gemini taught scientists many things about space flight as well as how people interact and work together. The third project was the Apollo project whose goal was to put a man on the moon. NASA achieved President Kennedy's goal when Neil Armstrong stepped on the moon on July 20, 1969. America became the new leader in the space race. Today, the United States continues to lead in the development of space technology and space exploration, but Russia and many other countries continue to be involved in space exploration as well.

In addition to manned space flight, the United States, the Soviet Union, and other countries worked to send scientific satellites into space to gather data about the earth including data on the land, atmosphere, and weather. Other satellites were developed for communications. The ECHO satellite, shown here, was NASA's first communications satellite, launched in 1960. This work continues today with many satellites being used for scientific purposes. One of the most famous scientific satellites is the Hubble Space Telescope.

Satellites are also used for commercial purposes such as television and telephone transmission. Recently, Global Positioning Satellites (GPS) and other navigational satellites have been sent up into space. Governments also send up spy or observation satellites to collect information on troops, arms, and other military data. Today, space satellites have become an integral part of our daily lives.

A third aspect of the space program is the development of unmanned space probes. These probes often travel to other planets that are too far away for manned visits. Most of the knowledge we have about the objects in our solar system has come from these probes. Initially, probes explored the closest object to earth—the moon. The *Luna*, *Range*, and *Surveyor* probes were sent to the moon to gather data before man was sent there. Later, probes were sent to explore the planets. The *Mariner 10* probe explored Mercury. The *Venera*, *Mariner*, and *Pioneer* probes visited Venus. And in the 1990s, the *Magellan* probe sent us the most accurate information we have from Venus.

Space Program

SPACE SATELLITE MODELS

Examine the pictures of the space probes and space satellites in this lesson. Then, design your own space satellite model using items around your house. For example, you could cover a Styrofoam ball with aluminum foil then attach toothpicks to make a Sputnik type of model, or you could form mod-

eling clay into the desired shapes and add tag board pieces to make the model. Space satellites come in all shapes and sizes depending on their purpose, so use your imagination. After completing the model, explain the function and purpose of your satellite to someone.

Optional—Model Rocket

For an exciting project, you can obtain a model rocket kit from a hobby store and build and launch your own model rockets. You can learn more about rocketry from the National Association of Rocketry at www.nar.org.

Mars has been explored by more space probes than any other planet. Over 30 probes, including the *Mariner* probes as well as *Viking 1* and *Viking 2*, have been sent to Mars. In recent years, attempts to send probes to Mars have failed. In 1998 and 1999, NASA lost two Mars probes, and in 2005 the European Space Agency's *Beagle 2* lander lost communication as it neared the planet. The most successful mission to Mars has been the *Spirit* and *Opportunity* rovers, which arrived on Mars in 2004 and transmitted data and pictures for over two years. Exploring Mars and understanding it is a high priority for NASA, and in 2004 President Bush announced an ambitious plan to send a manned mission to Mars, perhaps by 2030.

New Horizons probe

Most of the outer planets have been explored by one or more probes as well. *Pioneer 10* explored the asteroid belt and Jupiter. *Pioneer 11* and the *Cassini-Huygens* probe explored Saturn. And *Voyager 2* explored Jupiter in 1979, Saturn in 1981, Uranus in 1986, and Neptune in 1989. Pluto remains unexplored by space probes.

NASA has plans to continue exploring the planets. *Messenger* was launched in 2004 and should begin orbiting Mercury in 2011. The *New Horizons* probe (shown here) was launched in January 2006. It will take nearly 10 years to reach Pluto; this will be the first probe to explore this distant world. After flying by Pluto, *New Horizons* will then continue into the region beyond Pluto. Other possible NASA missions include a mission that would drop robotic landers into a crater at the south pole

WHAT DID WE LEARN?

- Who were the first people to talk about going into space?

- Who is considered the father of modern rocketry?

- What major event sparked interest in the development of the rocket for space travel?

- Who was one of the primary developers of rockets in the United States after World War II?

- What was the first man-made object to orbit the earth?

- Who was the first man in space?

- Who was the first American in space?

- Who was the first American to orbit the earth?

- Who was the first man to walk on the moon?

TAKING IT FURTHER

- Why are satellites an important part of space exploration?

- Why are space probes an important part of space exploration?

FUN FACT

Mike Melvill, the pilot of *SpaceShipOne*, was not the first private citizen to go into space. In April, 2001, millionaire Dennis Tito became the first space tourist. Tito paid Russia about $22 million for an eight day trip to the International Space Station. Since then there have been several space tourists on the space station. The first woman tourist in space was Anousheh Ansari, who spent 10 days on the space station in September 2006. If her name sounds familiar, it is because her family put up most of the money for the $10 million Ansari X prize.

of the moon and return samples to earth, and a mission that would orbit Jupiter from pole to pole.

Space exploration continues to fascinate humanity. We will probably never understand the broad expanse of the universe, but God has allowed us to glimpse its wonders through space travel. ■

COMMERCIAL SPACE FLIGHTS

Although government funded space programs have been the leaders in space exploration, privately funded space exploration got a giant boost on October 4, 2004. That is the day the *SpaceShipOne* completed its second flight into space in a two week period. But before we discuss this amazing machine, we must first talk about the X Prize.

Since the invention of aviation, privately funded prizes have spurred on new designs. Charles Lindbergh's famous flight across the Atlantic Ocean was in response to the Orteig Prize. In the 1920s Raymond Orteig offered a prize of $25,000 to the first person to fly non-stop across the Atlantic Ocean. In 1927, Charles Lindbergh was the first to successfully complete the trip and received the $25,000. In that same spirit, the Ansari X Prize was developed to spur research and development of privately funded space travel. The X Prize was started in May 1996, and consisted of $10,000,000 to be awarded to the first privately funded craft that could carry 3 people, fly to an altitude of at least 62.5 miles (100 km) and return to earth

safely and then repeat the trip again within a 14 day period. At least 7 teams signed up to compete for the prize.

The first ship to successfully fulfill the requirements was *SpaceShipOne* (shown here). Paul G. Allen, cofounder of Microsoft, provided most of the funding and Burt Rutan's company, Scaled Composites, designed and built the ship. Burt Rutan is the same man who designed the Voyager aircraft that successfully flew around the world without refueling in 1986. Mike Melvill was the pilot that flew the ship on both *SpaceShipOne* trips. The costs of putting *SpaceShipOne* into space is estimated at $25,000,000. Although the prize did not cover all of the costs of developing the ship, the publicity is expected to greatly boost the next step which is the development of *SpaceShipTwo*. The goal of *SpaceShipTwo* is to provide commercial flights into space for passengers and satellites.

SpaceShipOne is a **suborbital** ship, which means that it can go up into space and come back down but does not stay in space long enough to orbit the earth. A specially designed aircraft called

SpaceShipOne

the *White Knight* attaches to the top of *SpaceShipOne* to carry it to an altitude of about 50,000 feet. Then the space ship lights its rockets. *SpaceShipOne* uses a specially designed fuel that is a combination of a rubber-like substance and nitrous oxide. It uses these rockets to reach the altitude of about 62 miles, but it does not completely escape the earth's gravity. It would take about 30 times more energy to reach true orbit like the Space Shuttle does. Nevertheless, the future is bright for commercial space flights.

Suppose you were to design and build a reusable space craft. What would it look like? What would it be used for? Draw a picture of your space craft and explain how you think it would work.

Space Program

APOLLO PROGRAM

First flight to the moon

LESSON

30

What was the Apollo program and what did it accomplish?

Words to know:

Command Module

Service Module

Lunar Module

BEGINNERS

President John F. Kennedy challenged the scientists in America to send a man to the moon before the end of the 1960s, and they were able to do it. But it was not an easy thing to do. No one had ever been in space. So the first thing the scientists had to do was design a rocket and a space capsule that would send a man up into space and bring him back safely. This project was called the Mercury project.

Next, the scientists designed a space ship to take two men into space so they could work together on things that would need to be done on a mission to the moon. That was the Gemini project.

Once these missions were completed, NASA scientists and engineers had learned enough to begin the Apollo project. The Apollo project would take three men to the moon, land two of them on the moon, and then bring them all back to earth again. This was a very big project.

First a new and bigger rocket was needed to send the space ship all the way to the moon. This rocket was called the Saturn V (five) rocket. The Saturn V was really like three rockets in one. The first rocket lifted everything off of the ground. After the first rocket was fired, it was dropped off and the next rocket was fired. Finally the third rocket was fired and the space capsule was sent to the moon.

The space capsule had a place for the astronauts to sit and a place for the engines and computers. It also carried the Lunar Module, which was the vehicle that took Neil Armstrong and Buzz Aldrin down to the moon and back to the space capsule.

The first trip to the moon took place in July 1969. The astronauts took samples of moon rocks and moon dust and did many experiments. They were on the moon for a little more than two hours then returned to the space capsule. Astronauts made a total of six trips to the moon. The last trip to the moon was in 1975; however, in 2004 President George W. Bush authorized NASA to plan future missions to the moon, so someday people may return to the moon.

- What was the name of the project that sent men to the moon?

- How many stages or rocket engines did the Saturn V rocket have?

- Who were the first men on the moon?

To put a man on the moon and return him safely to earth was the challenge presented by President John F. Kennedy in 1961. He challenged the nation to reach this goal before the end of the decade. Against unbelievable odds, on July 20, 1969, Neil Armstrong became the first human being to walk on the surface of the moon.

In order to reach this goal, NASA, the National Aeronautics and Space Administration, first had to put a man in space, which they did with the Mercury project. Then they had to learn how to work in space, including working outside the capsule and docking with other space vehicles. All of these skills were needed for the trip to the moon. These goals were reached during ten Gemini missions. Finally, NASA was ready to reach for the moon, and the Apollo missions began.

Apollo required a completely different command and rocket system than what had previously been used. Going into orbit around the earth was a very different task than flying all the way to the moon and back. First, a new Command Module was designed. This cone-shaped module was the control center and living space for three astronauts. Attached to the Command Module was a Service Module containing an engine, fuel cells, and the power system for the Command Module. Together these two parts were called the CSM or Command and Service Module. The picture here is a model of the CSM.

A third module designed specifically for the Apollo missions was the Lunar Module. This was the vehicle used for landing on the moon's surface. It had two stages: a descent stage for landing on the moon and an ascent stage for taking off from the moon and returning to the Command Module. This picture shows a model of the LM or Lunar Module.

The CSM and LM were launched together into space by the giant Saturn V rocket. The Saturn V was a three-stage rocket that provided 7.5 million pounds of thrust. Each stage was a separate rocket engine that fired after the previous engine was extinguished. Stages one and

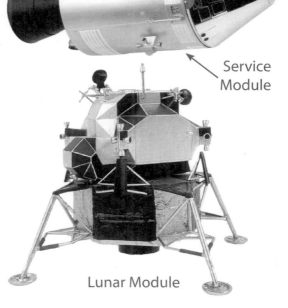

Command Module

Service Module

Lunar Module

two were used to get the modules into earth orbit. Each of these stages detached from the spacecraft after expending its fuel and fell back to earth. The third stage engine was used to propel the astronauts out of earth's orbit and to a lunar orbit around the moon. Once the fuel in the third stage was used up, it detached and remained floating in space.

Designing the Apollo space vehicle was only one part of getting men to the moon. The first several Apollo missions were training missions in earth orbit to prepare for the tasks that had to be accomplished on the moon. For example, the astronauts had to practice docking the lunar module with the command module in space so the astronauts could return from the moon.

 # TWO-STAGE ROCKET

The Saturn V was a three-stage rocket. Each stage was a separate engine. The first engine lifted the whole Apollo vehicle off the ground and into the air. The second stage propelled the astronauts into earth orbit. The third stage pushed the vehicle out of earth orbit and toward the moon.

Purpose: To make a model of a single-stage and a two-stage rocket

Materials: long string, soda straw, tape, two balloons

Procedure:

1. First, we will make a single stage rocket. Put one end of a long string through a soda straw then tape the string to a wall.

2. Tape the other end of the string to the wall on the opposite side of the room.

3. Next, blow up a balloon but do not tie the end. While holding the inflated balloon, tape it to the straw with the mouthpiece of the balloon pointing at one wall.

4. Pull the straw and balloon to the wall facing the mouthpiece of the balloon.

5. Release the balloon and watch it fly across the room.

6. Now we will make a two-stage rocket. Add a second straw to the string. Tape the second straw to the first straw.

7. Tape an inflated balloon to each straw as shown. Be sure that the balloon in the back is close enough to the front balloon to hold the mouthpiece of the front balloon shut. Rolling the mouthpiece of the front bal-

loon up will help to hold it in place.

8. Pull both balloons together toward the wall. Release the balloons. What happened?

Conclusion:

The single balloon moves the same way that a single-stage rocket does. The air pushes on the front of the balloon as it escapes out the back, propelling the balloon forward. With two balloons, the back balloon should push both balloons forward. Once the back balloon becomes smaller, the mouthpiece of the front balloon will be released and the front balloon should begin to deflate, thus increasing the speed of both balloons. This is similar to how a multi-stage rocket works on a spacecraft.

Single-stage rocket

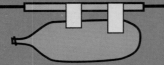

Two-stage rocket

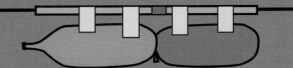

At first, it looked like Apollo would be a failure. The *Apollo 1* mission was a disaster with a fire breaking out before launch, killing all three astronauts inside the Command Module. This devastating accident forced NASA to redesign the Command Module, and the next several Apollo missions were unmanned ones designed to test every piece of equipment that would be used on the trip to the moon.

Finally, *Apollo 11* was ready to put a man on the moon. Three men, Neil Armstrong, Edwin (Buzz) Aldrin, and Michael Collins were selected to man this historic mission. After reaching lunar orbit, Collins remained in the Command Module and Armstrong and Aldrin were able to land on the moon. Neil Armstrong's famous line, "That's one small step for man, one giant leap for mankind," will forever remind us of that moment when man stepped beyond the limits of earth. You can hear a recording of this statement on NASA's web site at www.nasa.gov/mission_pages/apollo/apollo11_audio.html.

The two men remained on the surface of the moon for 2 hours and 13 minutes. They did several experiments, collected soil and rock samples, took pictures, and left an American flag and a memorial on the moon. Then they returned to the Command Module and the three astronauts safely returned to earth. NASA had accomplished what many people believed to be impossible.

Apollo 12 and Apollo missions 14–17 were also missions that put men on the moon. Astronauts on these missions performed additional experiments and explored different areas of the moon. Prior to the Apollo missions, no man had ever seen the far side of the moon. *Apollo 13* was an aborted mission when an explosion in the CSM damaged the module, requiring the astronauts to return home without stopping at the moon. The last Apollo mission was in 1975. It was a joint Apollo-Soyuz, American-Soviet project. There have been no manned missions to the moon since then, but in 2004 President George W. Bush announced plans for NASA to return to the moon by 2020. ■

FUN FACT

The longest time anyone has stayed on the moon was 74 hours 59 minutes during the Apollo 17 mission.

WHAT DID WE LEARN?

- What was the name of the NASA program whose goal was to put a man on the moon?
- What are the three modules in the Apollo spacecraft?
- What were the two parts of the lunar module designed to do?
- What was the name of the three-stage rocket used with the Apollo spacecraft?

TAKING IT FURTHER

- What is the advantage of a multi-stage rocket engine?

APOLLO 13

Overall, NASA has had an incredible track record of safety and success. Considering how little was known about space and how dangerous it is to attempt to work in the harsh environment of space, it is amazing that the Apollo program was so successful. However, one mission clearly brought home the delicate nature of working in space. *Apollo 11* and *Apollo 12* were successful missions with men visiting the moon and returning home safely. So *Apollo 13* was considered almost routine by many Americans who never expected there to be a problem.

The *Apollo 13* mission began fairly normally. There were some slight problems with the burn times of the second and third stage engines, but these were corrected and the astronauts began their journey toward the moon. They traveled for two days away from earth. Then suddenly things changed.

There was a loud noise inside the Command Module as an oxygen tank in the Service Module exploded. Suddenly, warning lights came on. Within three minutes only one of the three fuel cells had any power. As the astronauts checked the other systems they discovered that oxygen tank 2 had no oxygen and that oxygen tank 1

was slowly losing oxygen. This only left one oxygen tank unharmed. The astronauts knew immediately that they were in serious trouble. They knew they might not make it home.

The NASA engineers began examining the situation and working on solutions for getting the three astronauts back to earth before they ran out of oxygen and fuel. Because *Apollo 13* was most of the way to the moon, and

because of the shortage of fuel, it was decided that the best solution was to continue around the moon and to use the moon's gravity to give the space module a boost back toward the earth.

The Command Module was connected to the Lunar Module and the Lunar Module was unharmed by the accident. So to conserve energy, most systems were shut down in the Command Module and the men spent

Apollo 13 astronauts Fred Haise, John Swigert, and James Lovell are pictured during the press conference after their ill-fated mission.

Space Program

most of their time in the Lunar Module. This was not a perfect solution, because the Lunar Module was designed for only two people for two days. But three people would have to live there for four days before they could get back to earth.

The men took turns sleeping in the Command Module, but without energy the Command Module became very cold. By the time they got close to earth, the temperature in the Command Module had dropped to 38°F (3.3°C) and there was condensation over all the instruments, causing fear that something might short circuit when it was turned back on.

The astronauts had other problems to face as well. First, there was very little water available because the explosion had damaged the water tank and all three were suffering from dehydration. Also, the carbon dioxide level inside the Lunar and Command Modules was rising. Eventually, the astronauts were able to fashion an air filter out of miscellaneous parts inside the modules that helped get rid of the excess carbon dioxide.

The most frightening part of the mission may have been when *Apollo 13* had to go around the moon. At that time the astronauts were out of contact with earth for about 25 minutes. After they passed the back side of the moon and regained contact with earth, they did a short engine burn and began their trip home. As they began approaching the earth, the people in Houston realized that *Apollo 13* was off course because of the continued oxygen leak from tank 1. They had to plan another controlled firing of the engines to get the ship back on course so it would not miss the earth. As the ship got closer to earth, the astronauts used the Lunar Module engines, and made the necessary course correction.

When the ship was about nine hours from splashdown, the people in Houston decided there was enough power left to fully power up the Command Module, which helped to warm things up a little. Then the astronauts brought the necessary systems on line to make the landing. Finally, as they approached the earth, the astronauts disconnected the Lunar Module, which had been their life boat for the past four days. Then they jettisoned the Service Module. As it floated away their camera was able to take some pictures of the damage done by the explosion and the astronauts saw that a whole side of the module was missing.

The astronauts were able to complete the reentry procedure and splash down safely in the ocean. Their harrowing journey was over and they were safely back home. Although the ordeal of Apollo 13 shows the courage and ingenuity of man, it also shows the beauty and care that God took in designing our world, the only place we know of where life can exist.

FUN FACT

The memorial plaque that was left on the moon by the crew of *Apollo 11* reads:

HERE MEN FROM THE PLANET EARTH
FIRST SET FOOT UPON THE MOON
JULY 1969, A.D.
WE CAME IN PEACE
FOR ALL MANKIND

Space Program

THE SPACE SHUTTLE

Reusable parts

LESSON

31

How is the space shuttle designed and what is it used for?

BEGINNERS

After NASA sent several missions to the moon, it changed its focus to doing more work in space closer to home. In order to do this a new vehicle was needed, so NASA engineers designed the space shuttle. The space shuttle is a reusable space ship.

The space shuttle system has three parts: the orbiter, the liquid fuel tank, and the solid rocket boosters. When we think of the space shuttle, we usually think of the orbiter, which looks a lot like an airplane without jet engines. It has triangle-shaped wings and a tail fin. The second part of the system is the big orange fuel tank that contains most of the fuel used for takeoff. This is the only part of the system that does not get reused. The third part of the system is the solid rocket boosters that are attached on either side of the fuel tank at take-off. These rocket boosters contain additional solid fuel that is used to boost the shuttle into orbit around the earth.

The orbiter has a crew compartment, which is where the astronauts live while on a mission. It also has a payload bay behind the crew compartment where they store satellites or other equipment. This bay has a robotic arm that is very useful for launching satellites or for doing repairs to the space station or other activities in space.

Astronauts do many different things during space shuttle missions. Sometimes they do scientific experiments, other times they launch satellites, and many times they take supplies and people back and forth to the International

Space Station. When the mission is complete, the space shuttle returns to earth and lands in either Florida or California.

- **What is the space shuttle?**

- **What are the three parts of the space shuttle system?**

- **What piece of equipment is in the payload bay that is helpful to astronauts?**

After the last Apollo mission there was a six year period with no space flight by the United States. This was because a new type of space vehicle was being designed. Apollo had a huge rocket, the CSM, and the lunar module that together stood 363 feet high on the launch pad. Of this huge tower, only the CSM returned to earth. This was a very costly and inefficient design, so NASA set out to design a new space vehicle that could be reused. The result was the space shuttle.

The first space shuttle flight took place on April 12, 1981. This shuttle was called *Columbia*. Eventually four shuttles were built and quickly became the workhorses of the space program. The first four shuttles were named *Columbia*, *Challenger*, *Discovery*, and *Atlantis*.

There are three parts to the space shuttle system. First, there is the big orange fuel tank. This is the only part of the shuttle system that is not reusable. It holds the liquid fuel that is used for take-off. It is 153.8 feet (46.9 m) long, 27.6 feet (8.4 m) in diameter, and is the tallest part of the shuttle system. The second part of the shuttle system consists of the two solid rocket boosters. These attach on either side of the main fuel tank and are retrieved and reused after each launch. The final part of the shuttle system is the shuttle orbiter. As the name implies, the shuttle was designed to orbit the earth and cannot be used to fly to the moon. It orbits at about 150 miles (240 km) above the surface of the earth.

The shuttle orbiter is a delta, or triangular shaped, vehicle. It has a 78-foot (23.8 m) wingspan and is 122 feet (37.2 m) from nose to tail. The orbiter is much more spacious than the command module of the Apollo program. It has a crew cabin in the front with upper and lower decks. The upper deck is the flight deck. The lower deck has room for the astronauts to conduct experiments, eat, sleep, shower, and exercise. Behind the crew cabin is the payload bay. This large bay is often used to place satellites into orbit. It also contains a 50-foot (15 m) long robotic manipulator arm

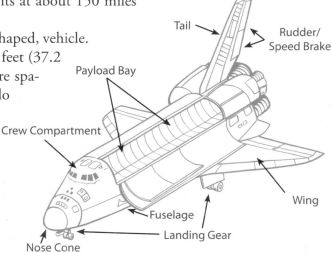

Tail

Rudder/ Speed Brake

Payload Bay

Crew Compartment

Fuselage

Wing

Landing Gear

Nose Cone

that can be used to move things around and to retrieve malfunctioning satellites. The space shuttle was used to place the Hubble Space Telescope into orbit (see picture on page 25). It was also used to repair the telescope when it did not work properly. Finally, the underside of the orbiter is covered with heat resistant tiles that protect the shuttle vehicle from the extreme temperatures experienced during reentry into the atmosphere.

Shuttle missions usually include seven crew members and last from 5 to 14 days. Missions are generally scientific in nature and include many experiments as well as launching satellites. Shuttle astronauts often perform EVAs, or extra-vehicular activities—better known as space walks. They often work in the payload bay to perform experiments or work on equipment. The shuttle is also used to ferry astronauts and materials back and forth to the International Space Station.

When a shuttle mission is complete, the orbiter normally lands in Florida, near Kennedy Space Center. Depending on the weather conditions in Florida, sometimes the shuttle lands in California. If this happens, it is flown back to Florida on the back of a specially modified 747 jet.

Just as with the Apollo project, there have been some heartbreaking disasters. On January 28, 1986, the space shuttle *Challenger* exploded shortly after take-off, killing all seven members aboard. The explosion was attributed to a leaking seal. After much redesign and testing, flights resumed in 1991. *Challenger* was replaced with a new shuttle named *Endeavor*. Flights continued relatively uninterrupted for over ten years. Then a second disaster occurred on February 1, 2003, when the shuttle *Columbia* burned up on reentry. NASA's three remaining space shuttles are slated for retirement by 2010 following the completion of the International Space Station (ISS). A new spacecraft—the Crew Exploration Vehicle (CEV)—has been tapped as its replacement, but is not expected to fly its first human-carrying mission until 2014. ■

FUN FACT

The original space shuttle was a prototype called *Enterprise*. This ship is now on display at the Smithsonian Air and Space Museum Annex. The *Enterprise* was used for tests beginning in 1977, but was never put into orbit around the earth.

THE SPACE SHUTTLE

Label the parts on the "Space Shuttle" worksheet.

WHAT DID WE LEARN?

- What is the main advantage of the space shuttle vehicle over all previous manned space vehicles?

- What are the main purposes of the shuttle program?

- What are the two main parts of the orbiter and what are their purposes?

TAKING IT FURTHER

- Why is the space shuttle called an orbiter?

- Why is the orbiter shaped like an airplane?

- Why does the orbiter have to be carried back to Florida if it lands in California?

Space Program

ORION

A possible program to replace the aging space shuttles is called the Constellation Program. NASA plans to replace the shuttle with a new Crew Exploration Vehicle that has been named *Orion*. The name *Orion* was chosen because Orion is one of the brightest and most easily identifiable constellations in the sky, and Orion has been used by sailors to discover new worlds in the past. NASA hopes that the new *Orion* will be instrumental in man exploring new worlds in the future.

Orion will be a space vehicle that serves a dual purpose. It will be able to take people and supplies to the space station or to the moon. The shuttle was never designed to go to the moon.

Conceptual drawings of *Orion* show it looking somewhat similar

Artist's rendering of the Orion crew exploration vehicle in lunar orbit

to the Apollo capsule, but with several important differences. First, *Orion* will be about 2½ times larger than the Apollo capsule so that it can hold up to six crew members. Also, it will have the latest computer technology and heat resistant systems. The space shuttle was designed to look something like an airplane, not because that shape is useful in space, but because that shape is

useful for landing on earth. A capsule cannot land on a runway, but must splash down into the ocean. This is less convenient than landing on a runway, but the capsule design is more heat resistant for high speed reentries that will occur on returning from the moon.

New rockets are also being designed to propel *Orion* into space. These rockets are called Ares I and Ares V. The name Ares was chosen because Ares is another name for Mars and the ultimate goal of the Constellation Program is to take people to Mars. These new rockets will take the liquid fuel engines and solid rocket boosters of the shuttle program to the next level.

The first manned flight of *Orion* is scheduled to occur no later than 2014. This will probably be a flight to the International Space Station. NASA hopes to send a man to the moon on the *Orion* no later than 2020. Check the NASA web site for the latest news on *Orion*.

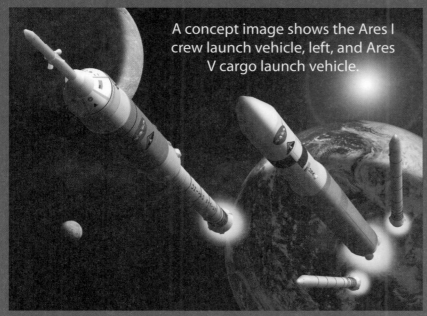

A concept image shows the Ares I crew launch vehicle, left, and Ares V cargo launch vehicle.

Space Program

Special FEATURE

RICK D. HUSBAND

1958–2003

The names of many astronauts are well known: Alan Shepherd, John Glenn, Neil Armstrong, Buzz Aldrin, and James Irwin to name a few. But the name of Rick Husband may never have become known if it weren't for the tragic events of February 1, 2003. Rick was the commander of the fateful shuttle that burned up on reentry, killing all seven crew members. Although many people will remember Husband for this event, the events of his life are much more important than the event that caused his death.

Rick Husband graduated from high school in 1975, and went on to earn undergraduate and masters degrees in mechanical engineering. Rick also joined the Air Force and became a test pilot and flight instructor. Rick dreamed of being an astronaut ever since he was a boy. But he was rejected by NASA three times. After the third rejection, Rick put his career in God's hands. His guiding verse was Proverbs 3:5–6, "Trust in the Lord with all your heart, and lean not on your own understanding; in all your ways acknowledge Him, and He shall direct your paths." He put his faith and his family as his top priorities and trusted in God to lead him. He decided to apply to the space program one more time and was accepted in 1994. After extensive training, Husband was assigned to be the pilot on the first shuttle mission to dock with the International Space Station in 1999.

Rick Husband was a highly-dedicated astronaut. He was very intelligent, determined, and well educated. Rick gave his best to his job. He was very skilled and was well respected by all who worked with him. However, Rick's devotion to his family and faith are what he would most want to be remembered for. Although his work was important to him, Rick remained true to his commitment to put God and his family first. Rick was a dedicated father and strong Christian man. He videotaped daily devotions for each of his children so they could continue their Bible studies while he was in space. While he and his wife were homeschooling their children, Rick was actively involved in a support group for homeschooling dads. He prayed with his crew before each mission and praised God openly from space.

Before his second shuttle mission, Rick was interviewed and said, "If I ended up at the end of my life having been an astronaut, but having sacrificed my family along the way or living my life in a way that didn't glorify God, then I would look back on it with great regret. Having become an astronaut would not really have mattered all that much." We should strive to live our lives to glorify God just as Rick Husband did. And we should remember this astronaut for his life of dedication to his faith and his family.

INTERNATIONAL SPACE STATION

Reaching for freedom

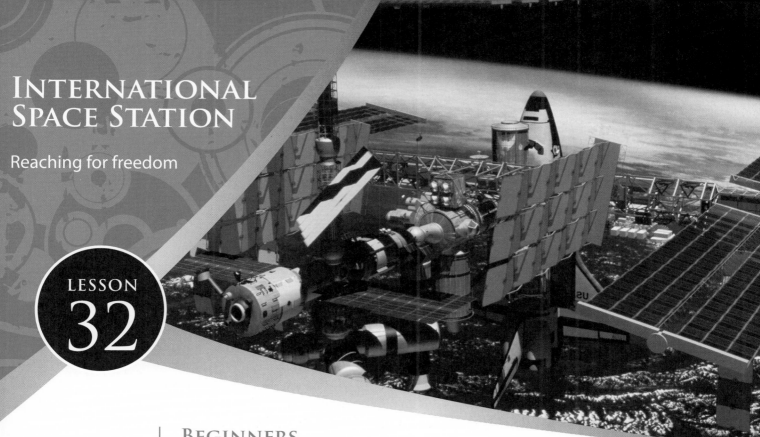

LESSON 32

Who built the International Space Station and how does it help us?

BEGINNERS

The space shuttle can be used to conduct experiments in space, but the space shuttle cannot stay in space for a long time. A more permanent space station was needed where people can live and work on experiments for months at a time. A new space station was started in 1998, and it is called the International Space Station. It was given the name International because several countries are working together to build and operate the space station. Although the United States is one of the main countries building the space station, Russia, Canada, Japan, and several nations in Europe are also helping.

The space station has several parts. The most obvious feature is the solar panels that help provide energy for the space station. There are crew quarters and several laboratories. There are usually three astronauts living on the space station at a time. Most of the astronauts have been from either the United States or from Russia, but several have been from other countries as well.

On the space station, astronauts can do experiments that cannot be done on earth because there is very little gravity on the space station. They experiment with plants and animals, with chemicals, and with medicines. Many of the things the astronauts have learned have benefited us here on earth.

- What is the name of the current space station?

- How does the space station get power?

- Why can astronauts do different experiments on the space station than on earth?

NASA quit conducting flights to the moon after the Apollo missions were over because flights to the moon are very costly, and people at NASA believed they had learned what they could from manned flights there. Instead, NASA has concentrated on shuttle flights for scientific purposes. However, most shuttle flights are 14 days or less in duration. The United States and other countries, especially Russia, have also seen the benefits of having astronauts in space for extended periods of time to conduct long-term research in a zero or micro-gravity environment.

To provide a place to conduct these long-term experiments, both the U.S. and Russia have placed space stations in orbit around the earth. The first space station, called *Salyut*, was launched in 1971 by the Soviet Union. The United States launched its first space station, *Sky Lab*, in 1973. These stations were used to conduct many experiments, both for civilian and military purposes. The orbits of both of these stations deteriorated and they burned up on reentry.

Backdropped by the blackness of space and earth's horizon, the International Space Station is seen from Space Shuttle *Discovery*.

Space Program

In 1986 the Soviets launched a new space station called *Mir. Mir* was the first permanently-manned space station. It operated for 15 years, and astronauts from dozens of nations visited and worked there. Various modules were added and sometimes removed depending on the required functions of the station. What was learned on *Mir* provided stepping stones for the next stage in space stations.

In January 1984 U.S. President Ronald Reagan announced that NASA was beginning a new space station program that would be international in scope. This new space station was to be called *Freedom*, and was to be a joint effort between the United States, Canada, Japan, the European Space Agency, and other friendly countries. Work began on the design of space station *Freedom*. Then, after the fall of the Soviet Union in 1989, relations improved between the United States and Russia. And in 1991, President George Bush worked to include Russia in the planning of the space station. With the addition of Russia, the name was changed to the International Space Station.

Construction on *Freedom* was originally scheduled to begin in 1995, but due to many factors, the International Space Station was actually begun in 1998. It was estimated that it would be completed in 8 years. However, shuttle flights were canceled after the *Columbia* explosion in 2003. Several parts of the shuttle were redesigned

WATER BALLS

One interesting observation in space is that water, which on earth conforms to whatever container it is in, actually forms balls of water in zero gravity. We of course can't observe this phenomenon on earth, but we can get an idea of how it occurs by doing the following.

Purpose: To see why water forms balls in space

Materials: water, waxed paper, toothpick or knife

Procedure:

1. Place a few drops of water about 2 inches apart on a piece of waxed paper. Note how these drops form into bubble shapes instead of flowing across the paper.

2. Using a toothpick or the edge of a butter knife, slowly push one drop of water toward another. What happens when the drops get close together?

3. Try to separate a larger drop of water into two separate drops. How difficult is this?

4. Slowly move a wet toothpick near the drop of water. What happens when it gets near?

Conclusion:

Water molecules have an attraction for each other, and one drop is pulled toward the other. In space, this attraction causes water to gather into balls, since gravity is not pulling them flat onto a surface like on earth. In our experiment, when the drops were moved closer together, they were attracted to one another. When you tried to separate them, the attraction of the water molecules prevented the larger drop from separating easily. The water drop on the paper is attracted to the water on the toothpick so the drop moves toward the toothpick even before the toothpick touches it.

and flights resumed in July, 2005. This has delayed the completion date for the station until 2010; however, the station is still functional and astronauts have been conducting useful research there continuously since 2001.

The International Space Station is being built in stages, with U.S. space shuttles and Russian Soyuz rockets launching pieces into space where they are then joined together by astronauts during space walks. Enough of the space station is built that it is now habitable and being used to perform many important experiments. Astronauts moved in on November 2, 2000, and are currently utilizing seven different research facilities. There are usually 3 crew members in the station for several months at a time. The average stay is between 128 and 195 days. Most of the occupants have been Americans and Russians; however, there have been astronauts from many other countries working in the station as well.

The crew of the space station conducts many experiments that cannot be done on earth. Some of these involve growing crystals. Crystals form more perfectly without gravity. These crystals are being used to make new medicines. Other experiments involve growing plants in micro-gravity and testing the long-term effects of micro-gravity on plants as well as people. One experiment tested the long-term effects of space walks on the lungs. Liver tissues are grown in this environment in hopes of finding better screening for patients prior to transplants. Also, various energy experiments are conducted to find new sources of energy. These, and many other experiments, hold the promise of great advancements in science in the future. Also, over 30,000 pictures have been taken of the planet and sent back to earth. The research done at the International Space Station is expected to yield many other useful results for society in the future. ■

WHAT DID WE LEARN?

- What is the International Space Station?
- Why do countries feel there is a need for a space station?

TAKING IT FURTHER

- What shape would you expect a flame to be on the space station?

ASSEMBLING THE SPACE STATION

Assembling the space station is a monumental task. Most items that go into space such as satellites and space shuttles are assembled on earth and then transported into space; however, the space station is too big to be done that way. It has been assembled and is continuing to be built piece by piece in space. Assembling a ship or space station in space has many challenges that are not real considerations on earth. What problems or challenges can you see with assembling a large space ship in space?

The first problem faced by astronauts is the lack of gravity. Because there is no gravity, everything must be connected to something else or it may drift away forever. Moving in space is very different from moving on earth so astronauts must train for this special environment. There is one advantage to a low-gravity environment, however. Heavy pieces of the space station, weighing as much as 35,000 pounds, would be very awkward to move on earth, but can be moved relatively easily in space because of their near weightlessness.

The second problem is no atmosphere. All assembly work must be done either by robotic arms or by space walks in space suits. The space station has a robotic arm that can be moved along a truss system to place it where it is needed. In addition, the space shuttle has a robotic arm in its cargo bay. These two arms can be used together to put pieces into place. Other times space walks are necessary. Astronauts must leave the space station or the space shuttle in bulky space suits to complete the connections and installation of the new parts for the space station, as pictured above. These are problems that do not have to be dealt with on earth. This makes assembly difficult and time consuming.

A third problem is related to the first two. Because there is no other source for power and oxygen, all assembly work must be done without disrupting the power and life support inside the space station. When new power systems are added, they must be wired into the existing station without disrupting the power that is already in place. Mike Suffredini, the NASA station program manager said, "It's like building a ship in the middle of the ocean from the keel up. You've got to float and you've got to sail. All this has to occur while you're actually building the ship, and that's what the station is like."

Astronauts are up to the challenge, however. They train for months before a mission so that they can assemble the new pieces and add to the functionality of the station. So far, the space station has been a success, and many important experiments are being conducted there. In addition, what scientists have learned from building the space station will help them when they eventually build a permanent base on the moon and perhaps someday on Mars.

ASTRONAUTS

Modern day explorers

LESSON

33

BEGINNERS

Have you thought about becoming an astronaut? Does that sound exciting to you? If you want to be an astronaut, you have to learn math and science, and you have to be in good shape. Then, when you are an adult, you can apply to the space program. Many people apply but only a few are chosen to become astronauts.

Once a person is chosen to become an astronaut, he or she must go through a long training period. Astronauts must learn how to work in an environment without gravity. They must also learn to operate the equipment that they will be using. This could mean learning to fly the space shuttle, or learning to install new equipment on the space station. Other astronauts conduct scientific experiments in space and must train to do those experiments just right.

Astronauts also have to learn to work while wearing space suits. Much of the work that is done in space can be done inside the space shuttle or the space station where there is air to breathe and air pressure, but sometimes work has to be done in space where there is no air and no air pressure. Then, astronauts have to wear special suits that provide all the things needed to keep them alive. This includes air and pressure. The space suits also provide heat and cooling, water, and protection from radiation. Space suits also have communication equipment so the astronauts can talk with each other and with the people back on earth. Sometimes astronauts put on rocket packs that help them move around in space.

You will have to work very hard if you want to be an astronaut, but it will be a very rewarding job if you make it.

- What are two important subjects to study in school if you want to become an astronaut?

- What things do space suits provide that are missing in space?

- How does an astronaut move around in space?

Many children dream of being astronauts when they grow up. The excitement and adventure of traveling into space is very appealing to some. But how does a person become an astronaut?

A few traits are common to all astronauts. First, all astronauts have an education in math and science, usually in engineering, physics, chemistry, or some other technical field. So if you want to be an astronaut, you need to work hard in math and science. Also, physical fitness is very important. Your body must be in good shape if you want to be an astronaut.

Many thousands of hopeful people apply to the space program each year, but only a handful are selected to become astronauts. After being selected, an astronaut-to-be must go through several years of training before he or she is ready to go into space.

Astronauts must learn to work and live in a micro-gravity environment. This may not seem hard, but imagine trying to do a task when your tools try to float away, what you are working on tries to float away, and you try to float away, all in different directions. Even taking a shower becomes difficult without gravity to pull the water down. One way astronauts prepare for working in a weightless environment is to work in a giant swimming pool. Working in a spacesuit under water simulates working in space.

A second way that astronauts train for zero gravity is to fly in a special jet. This jet climbs to about 36,000 feet (11,000 m) and then quickly dives to about 24,000 feet (7,300 m), allowing the passengers to experience weightlessness for 20–30 seconds (shown here). They repeat this up to 40 times in one day. This often makes astronauts feel sick, so this jet is nicknamed the *Vomit Comet*.

In addition to physical training, astronauts must learn all about the controls of the space shuttle and how to perform

FUN FACT

Space Camp in Huntsville, Alabama is a program for children that allows them to see what it is like to be an astronaut. Participants get to try astronaut water survival, launch model rockets, and complete a simulated space shuttle mission. Some participants design experiments that actually go on future shuttle missions. For more information on Space Camp, visit their website at www.spacecamp.com.

SPACE SUITS

Purpose: To simulate an astronaut in his space suit

Materials: winter clothing (hat, gloves, coat, snow pants, boots), hand mirror, building blocks, nut and bolt, bicycle helmet or motorcycle helmet with face mask (if available)

Procedure:

1. Pretend to be an astronaut by dressing in a full set of winter clothes, including snow pants, a coat, gloves, bulky boots, and a helmet or winter hat.

2. Now that you are fully dressed, try to perform various tasks such as building with blocks, exercising, and screwing a nut onto a bolt.

3. If you are wearing a motorcycle helmet, breathe hard inside the helmet for a few seconds. If no helmet is available, breathe onto a hand mirror. What happened?

Questions:

 • How difficult was each task?

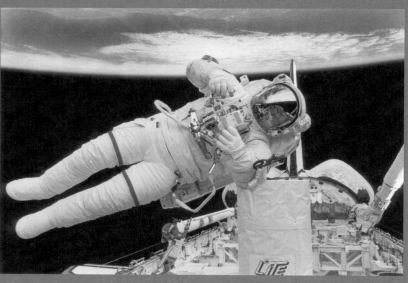

 • What problems did you encounter?

Conclusion:

The snow pants are similar to the pants that astronauts wear. A coat is similar to the top half of the space suit. Astronauts wear bulky boots and gloves as well. A motorcycle helmet with a face mask is similar to an astronaut's helmet.

As you can tell, working would be difficult in a space suit. Astronauts must train to do their jobs in their full space suits. They often use tools that are specially designed for easy gripping by gloved hands.

Condensation appeared inside the mask or on the mirror. This is a problem that space suits have been designed to handle. Also, did you get hot while wearing all the extra clothing? Space suits are equipped to cool the astronauts when in sunlight and to warm them when in darkness.

whatever experiments are required. They must understand physics and astronomy very well in order to do their jobs.

In order to survive in space, astronauts must wear space suits when not inside a space shuttle or space station. These suits are designed to protect the astronauts from the extreme heat and cold of space; to provide pressure, oxygen to breathe, and water to drink; and to protect them from the strong radiation found in space. These are all functions that are performed by our atmosphere on earth. Also, the space suit has communication equipment so astronauts can talk to each other. Finally, a space suit can be fitted with an EMU, Extravehicular Mobility Unit, which is a rocket pack that propels the astronaut through space. The EMU is especially helpful when working outside the space shuttle. The astronaut shown at left is using the EMU outside the space shuttle.

You don't have to be an astronaut to be part of the space program. The vast majority of the people involved in the space program never go into space. They include launch support staff, technical support staff, manufacturers, and teachers. Thousands of people are involved in every space launch. They are designing and building each piece of equipment, programming computers, training astronauts, planning missions, and controlling every detail required to make the mission a success. So even if you are not an astronaut, you can be involved in the space program. ■

WHAT DID WE LEARN?

- What are some ways that astronauts train for their missions?
- What conditions in space require astronauts to need space suits?

TAKING IT FURTHER

- What are some things you can do if you want to become an astronaut?
- What would you like to do if you were involved in the space program?

RESEARCH AN ASTRONAUT

There have been hundreds of astronauts since the space program began. At first the astronauts were all from the military, primarily Air Force pilots. Today astronauts can be civilians, although most are still from the military. Early astronauts were chosen for their flying ability, but today, many astronauts are chosen for their scientific backgrounds in many areas.

Choose an astronaut to research. Find out about that person's background, early career, and achievements as an astronaut. You can choose any astronaut you are interested in, but in the list at right are the names of a few of the most famous ones.

Present your research to your family, class, or other group.

John Glenn
James Lovell
Alan Shepherd
Michael Collins
Neil Armstrong
Virgil 'Gus' Grissom
Scott Carpenter
Judith Resnik
Sally Ride
James Irwin
Eileen Collins
Mae Jemison

SALLY RIDE

1951–PRESENT

On June 18, 1983, Sally Ride made history. She became the first American woman in space. She also became the youngest American to orbit the earth. Sally was part of the five-person crew whose mission was to launch two space satellites from the *Challenger's* payload bay. Sally was able to operate the 50-foot (15 m) robot arm to place the satellites in orbit. This was certainly one of the most memorable experiences of Sally Ride's life.

Sally Ride was born on May 26, 1951, in Los Angeles, California. Her father, Dr. Dale Ride, was the assistant to the president of Santa Monica College and an elder in the Encino Presbyterian Church. Her mother, Joyce Ride, stayed home to raise Sally and her younger sister Karen. The Rides encouraged both of their daughters to work hard and always do their best, and they were proud of both of them.

Growing up, Sally was very good at sports, especially tennis. She attended the Westlake School for Girls and worked hard to graduate a year early near the top of her class. Her favorite subject in school was science and when she attended Stanford University, she earned a double major in English

and Astrophysics. Before completing college, Sally took a year off from school to pursue a career as a professional tennis player. After deciding that she could not compete as a professional, Sally returned to school and completed her undergraduate degree. She continued her studies and received an MS and a PhD in Astrophysics.

While attending Stanford, Sally read a magazine article saying that NASA was looking for astronauts. Sally had a strong desire to become an astronaut and applied to the space program in 1978, along with 8,900 other applicants. She was one of only 35 people accepted into the program that year. Five other women were also among those 35 chosen to become astronauts. After being accepted, Sally moved from California to Houston, Texas to begin her training at Johnson Space Center. There she studied flight engineering and became a pilot. She had to work very hard, both physically and mentally.

The year 1982 was a very special year for Sally Ride. After four years of hard work at the space center, Sally was chosen to become the first American woman in space. But 1982 held other joys as well. On July 24, 1982 Sally married Dr. Steven Hawley. Sally's sister

Karen had become a minister and Steven Hawley's father was also a minister, so they jointly performed the wedding ceremony.

After being chosen for a 1983 *Challenger* mission, Sally had to begin training for the mission. For over a year, she practiced all of the activities that she would perform in space, including training in simulators, practicing emergency procedures, working in spacesuits underwater, and riding in the *Vomit Comet*—a jet that allows the passengers to experience true weightlessness for short periods of time.

Finally, the crew was ready. The commander was Robert Crippin, the pilot was Rick Hauck, and the mission specialists were John Fabian, Norman Thagard, and Sally Ride. The shuttle lifted off on June 18, 1983 for the six-day mission. All of the years of study and training finally paid off as Sally reached her goal of flying into space.

Sally was able to fly on a second mission in *Challenger* in 1984. This was an eight-day mission, and the main scientific goal was to study the earth's atmosphere. The shuttle crew used high-resolution cameras and radar to study the earth. Sally was preparing to go on a third mission when the *Challenger* exploded shortly after take-off in 1986. After this accident, Sally Ride was appointed to the Presidential Commission in charge of investigating the cause of the accident.

After completing her work on the commission, Sally retired from NASA in 1987. She accepted a position as a Science Fellow at the Center for International Security & Arms Control at Stanford University. In 1989 Dr. Ride was named Director of the California Space Institute & Professor of Physics at the University of California, San Diego.

In addition to teaching, Sally is very interested in helping girls who are interested in science and math to pursue these fields. To help encourage girls to study technical areas, Sally has started an organization called Imaginary Lines. This organization sponsors the Sally Ride Science Club and other activities that encourage girls to study math, science, and technology. Sally has also written four children's books about space.

Sally Ride is a very private person and does not like a lot of attention. She does not seek fame or fortune. She is described as adventuresome, tough, and athletic; she also has a good sense of humor and enjoys writing and teaching. But she will always be known as the first U.S. woman to go into space.

SOLAR SYSTEM MODEL: FINAL PROJECT

Showing what's out there

LESSON 34

Build your own model of the solar system.

Recall from lesson 2 that models are helpful for allowing us to visualize something that is either too big or too small to easily see. You can demonstrate your understanding of the solar system by building your own model. Copernicus, Galileo, and others showed that the sun is the center of our solar system and the planets revolve around it. At one time it was believed that there were five planets besides the earth revolving around the sun. With the invention of the telescope and better lenses, astronomers have detected a total of eight planets revolving around the sun, as well as other smaller objects such as asteroids and comets.

Review the order of the planets by singing the song you learned in lesson 11. You can also remember their order by learning the sentence, "My Very Excellent Mother Just Served Us Nachos" (Mercury, Venus, Earth, Mars, Jupiter, Saturn, Uranus, Neptune). Now you are ready for the fun of building your own model. ■

Jan. 15

FINAL PROJECT: SOLAR SYSTEM MODEL

Purpose: To assemble a model of the solar system

Materials: nine Styrofoam balls, two Styrofoam rings, paint, craft wire (see chart below for details)

Procedure:

1. Paint each Styrofoam ball and ring as follows and allow them to dry. If you have a book available with photos of the planets, you can use the pictures as guides for how to paint each planet, or use the pictures in this book.

2. After all of the pieces are dry, place the sun on the base.

3. Put Saturn's rings around Saturn.

4. Attach each planet to the sun using stiff craft wire cut to the following lengths:
   Mercury—2½ inches
   Venus—4 inches
   Earth—5 inches
   Mars—6 inches
   Jupiter—7 inches
   Saturn—8 inches
   Uranus—10 inches
   Neptune—11½ inches

5. If you want to, make small moons out of modeling clay and attach them to their planets with small pieces of wire.

6. You can also add an asteroid belt by gluing several small pieces of clay or pebbles together and then gluing them to a piece of wire 6½ inches long and placing them so that they are between Mars and Jupiter.

When your model is complete, share what you learned with your class or family.

Ball/Ring Size	Represents	Suggested Color
4½-inch ring	Base of the model	Any color you like
5-inch ball	Sun	Yellow
1¼-inch ball	Mercury	Red/brown
1½-inch ball	Venus	Reddish yellow
1½-inch ball	Earth	Blue with green/brown continents
1¼-inch ball	Mars	Red
4-inch ball	Jupiter	Red/orange—Don't forget the Great Red Spot
3-inch ball	Saturn	Peach
4½-inch ring	Saturn's rings	Striped—any colors you like
2½-inch ball	Uranus	Blue/green
2-inch ball	Neptune	Blue—May include a Dark Spot

WHAT DID WE LEARN?

- What holds all of the planets in orbit around the sun?

- What other items are in our solar system that are not included in your model?

TAKING IT FURTHER

- Why do the planets orbit the sun and not the earth?

PLANET STATISTICS

For each planet, make a card that can be displayed below or in front of the planet. An index card can be used for this. On each card include the name of the planet and the statistics from the Planet Statistics chart on page 119. Include any other interesting information you have learned about each planet.

FUN FACT

The suggested lengths for the wires are convenient for making a model but are not very accurate for the actual size of each planet's orbit. If the sun's diameter were 5 inches (12.7 cm), like in our model, then the planets would actually be the following distances away from the sun:

Mercury—17 feet (5.2 m)
Venus—32 feet (9.8 m)
Earth—44.5 feet (13.6 m)
Mars—69 feet (21 m)
Jupiter—232 feet (70.7 m)
Saturn—425 feet (129.5 m)
Uranus—856 feet (261 m)
Neptune—1,341 feet (408.7 m)

And remember, the sun is a close star. Other stars are much, much farther away. Hopefully, this gives you a glimpse of just how big space really is.

Space Program

CONCLUSION

Reflecting on our universe

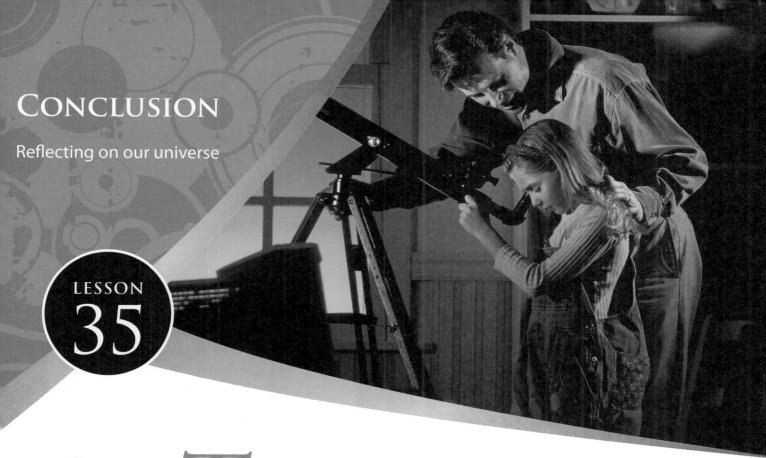

LESSON 35

Reflecting on our amazing universe

The universe is so vast that our minds cannot truly comprehend how big it is. The universe gives us just a glimpse into the awesome power of the God who created everything in it. Think about each thing that God created: stars, planets, moons, comets, asteroids, quasars, nebulae, our sun, and best of all, our planet earth. God loves us so much that He created a beautiful place for us to live and put it in a universe we can only begin to understand. ■

REFLECTING ON GOD'S WONDERFUL CREATION

Purpose: To reflect on God's creation and His goodness

Materials: blanket, flashlight, Bible

Procedure:

1. Take a blanket, a flashlight, and your Bible, and go outside on a clear night to observe the heavens and reflect on God's mighty power.

2. Spread the blanket on the ground in an area where there is not too much light. The darker the area, the better your view will be of the stars.

3. Use the flashlight to read the following Scripture passages. Reflect on each passage and thank God for His goodness as you look at the stars.

Psalm 8:1–9
Psalm 19:1–6
Psalm 102:25–28
Psalm 108:3–5
Psalm 119:89–90
Psalm 139:7–10

Note: this activity can be done indoors, looking out a window if weather prevents viewing the stars outside, but is not as much fun.

Space Program

WHAT DID WE LEARN?

- What is the best thing you learned about our universe?

TAKING IT FURTHER

- At the beginning of this book you wrote down some questions you had about astronomy. Check your list and see if you have found out the answers. If not, find a book or online resource for the answers.

- What would you like to learn more about? (Visit your library or search online for more information.)

- To study more about astronomy and to see how the universe declares the glory of God, go to www.answersingenesis.org/go/astronomy.

Space Program

GLOSSARY

Asterism   An "unofficial" constellation or group of stars

Asteroid   A relatively small rock in a regular orbit around the sun

Astronomy   The study of the planets, moons, stars, and other objects in space

Aurora australis   The southern lights

Aurora borealis   The northern lights

Big bang theory   All that exists in our universe came from a cosmic explosion about 14 billion years ago

Chromosphere   Sun's atmosphere closest to the sun lying just above the photosphere

Comet   A core of ice in a regular orbit around the sun

Command Module   Apollo capsule housing the astronauts

Convective zone   Outermost layer of the sun's interior where convection currents occur

Core   Center of the sun where the thermonuclear reaction takes place

Corona   Sun's atmosphere that spreads out into space

Crescent moon   Less than ½ of the moon is lit

Equinox   First day of spring or autumn

Escape velocity   Speed required to overcome gravity

Full moon   Full circle of the moon is lit

Geocentric model   Everything revolves around the earth

Gibbous moon   At least ½ but not all of the near side of the moon is lit

Gravity   The force one body exerts on another due to its mass

Heliocentric model   Everything in our solar system revolves around the sun

Inferior planets   Planets with orbits inside or closer to the sun than the earth's orbit

Jovian   "Jupiter-like," planet made of gas

Light-year   The distance light travels in one year

Lunar Module   Vehicle designed to be used on the moon

Lunar eclipse   When the earth is directly between the sun and the moon and earth's shadow falls on the moon

Maria   Basalt-filled plains on the moon's surface

Meteor   Rock or other debris pulled from space into the earth's atmosphere by gravity

Meteorite   A meteor that reaches the surface of the earth

Meteoroid   A small particle of rock or other debris in space

Milky Way   The galaxy to which our solar system belongs

NASA   National Aeronautics and Space Administration

Nebula   A cloud of gas and dust in space

New moon   None of the near side of the moon is lit by the sun

Newton's law of gravitation   Any two objects exert a pull on each other that is proportional to the product of their mass, and inversely proportional to the square of the distance between them

Nova   A star experiencing an explosion

Photosphere   The visible (light-emitting) surface of the sun

Radiative zone   Second layer of the sun's interior where energy moves outward from the core as electromagnetic waves

Retrograde motion   Movement that is backwards or opposite that of earth

Revolution   Movement of a heavenly body around another heavenly body

Rotation   Movement of a heavenly body around its axis

Satellite   Something that orbits another body

Service Module   Module attached to Command Module housing supplies and equipment

Solar eclipse   When the moon is directly between the sun and the earth, the moon casts its shadow upon earth, blocking the sun

Solar energy   Energy from the sun

Solstice   First day of summer or winter

Space probe   Man-made instrument designed to explore objects in space

Sunspot   Dark cooler area on the surface of the sun

Super nova   A star experiencing an unusually large explosion

Superior planets   Planets with orbits outside or farther from the sun than the earth's orbit

Synchronous orbit   An orbit of a satellite around a rotating body, such that one orbit is completed in the time it takes for the body to make one revolution

Terrestrial   "Earth-like," planet of solid rock

Waning   Becoming smaller

Waxing   Growing or getting bigger

CHALLENGE GLOSSARY

Aphelion   Location in orbit furthest from the sun

Celestial equator   Projection of the equator onto the night sky

Centripetal force   Force that causes something to move in a circle

Dark side of the moon   Side facing away from the sun

Degrees of declination   Location above or below the celestial equator

Ellipse   A squashed circle shape

Far side of the moon   Side facing away from the earth

Foucault pendulum   A pendulum that swings independent of the movement of the earth

Hours of ascension   Location right of the prime hour circle

Interferometry   Combining light from two telescopes into one image

Light side of the moon   Side facing the sun

NACA   National Advisory Committee on Aeronautics

Near side of the moon   Side facing the earth

Penumbra   The outer, lighter region of a sunspot or shadow

Perihelion   Location in orbit closest to the sun

Prime hour circle   Projection of the prime meridian onto the sky

Rays   Bright streaks of lunar material radiating from craters

Rills   Valleys on the moon

Shepherding moons   Moons around Saturn (or another planet) that affect the location of the rings

Suborbital   Going into space and coming back without orbiting the earth

Supersonic   Faster than the speed of sound

Umbra   The inner dark region of a sunspot or shadow

Vernal equinox   Point that celestial equator and prime hour circle meet; the sun is here on the first day of spring

Vortex   A spiraling mass of water or air

INDEX